TRANSACTIONS
of the
AMERICAN PHILOSOPHICAL SOCIETY
Held at Philadelphia
For Promoting Useful Knowledge
Volume 92, Part 5

ADDITIONS TO THE PLEISTOCENE MAMMAL FAUNAS OF SOUTH CAROLINA, NORTH CAROLINA, AND GEORGIA

Albert E. Sanders

American Philosophical Society
Philadelphia • 2002

ISBN: 0-87169-925-7
US ISSN: 0065-9746

Library of Congress Cataloging-in-Publication Data

Sanders, Albert E.
   Additions to the Pleistocene mammal faunas of South Carolina, North Carolina, and Georgia / Albert E. Sanders.
        p.   cm. — (Transactions of the American Philosophical Society; v. 92, pt. 5)
   Includes bibliographical references and index.
   ISBN 0-87169-925-7 (pbk.)
      1. Mammals, Fossil—South Carolina. 2. Mammals, Fossil—North Carolina. 3. Mammals, Fossil—Georgia. 4. Paleontology—Pleistocene. I. Title. II. Series.

QE881.S26 2002
569'.0975—dc21   2002038601

TABLE OF CONTENTS

ABSTRACT

Discoveries of vertebrate fossils near Charleston and Myrtle Beach, South Carolina, and in Brunswick County, North Carolina, have provided new records of 37 Pleistocene mammal taxa on the Atlantic Coastal Plain. Two North Carolina specimens of early Irvingtonian age are the first records of the genus *Neofiber diluvianus*-like animals south of Pennsylvania and are the oldest evidence of the genus *Neofiber*. Additional records of *Neofiber alleni* True from South Carolina also are included. The holotype molar of *Neochoerus pinckneyi* Hay is found to be of Mid-Pliocene age instead of Pleistocene, and the holotype premolar of *Alces runnymedensis* Hay is referred to *Cervalces*. Early Pleistocene remains of *Dasypus*, *Eremotherium*, *Equus*, *Cervus*, and *Hydrochoerus* in South Carolina are documented, along with new records of *Tremarctos floridanus* (Gidley), *Arctodus pristinus* Leidy, *Smilodon fatalis* (Leidy), *Castor canadensis* and *Mammut americanum* (Kerr) from South Carolina. The first fossil records of *Erethizon dorsatum* Linnaeus, *Panthera leo atrox* (Leidy), *Puma concolor* Linnaeus, *Erignathus barbatus* Gill, *Hemiauchenia* cf. *H. macrocephala* (Cope), *Cuvieronius* Osborn, and *Monachus tropicalis* (Gray) in South Carolina are reported, as well as the first record of *Monachus* from Georgia. A mandibular ramus tentatively referred to *Puma concolor* by Allen (1926) is redetermined as *Miracinonyx inexpectatus* and is the first evidence of cheetah-like cats in the fossil record of South Carolina. A well preserved right mandibular ramus from Charleston County is designated as the neotype of *Arctodus pristinus* Leidy, 1854. The boundary of the Irvingtonian/ Rancholabrean Land Mammal Age is placed at 0.24 Ma on the basis of a *Bison* astragalus from the Ten Mile Hill Beds (Late Middle Pleistocene).

ACKNOWLEDGMENTS

I am indebted to Robert E. Weems of the U.S. Geological Survey for many useful discussions of the Pleistocene stratigraphy of the Lower Coastal Plain of South Carolina and to Laurel Bybell of the USGS for advice about nannoplankton dates. My sincere gratitude goes to Barry Albright, C.B. Berry, Robert Johnson, Gracie Marvin, Bill Palmer, and Gary Towles for their generous donations of specimens to The Charleston Museum in connection with the present study. For past donations of specimens I thank Mike Crowell, Edmund R. Cuthbert, Jr., Doris Holt, Vance McCollum, and Margaret Weeks. For the loan of specimens in their care I am especially grateful to Edward Daeschler (Academy of Natural Sciences of Philadelphia), S. David Webb (Florida Museum of Natural History), James L. Knight (South Carolina State Museum), Robert Purdy and David Bohaska (U.S. National Museum of Natural History), Charles R. Schaff (Museum of Comparative Zoology, Harvard University), and Richard E. Tedford (American Museum of Natural History). Barry Albright lent valuable assistance in locating certain specimens in the Florida Museum of Natural History collection. For the loan of specimens in their private collections I thank Ray and Bettie Harkless, Lee Hudson, Chet Kirby, Terry Lee, Ned Riddle, and the late Don Marvin. For valuable discussions and advice about the status of various Pleistocene taxa, and, in some cases, assistance in the identification of specimens, I express my sincere appreciation to Fred Grady, Dale Guthrie, C. R. Harington, Richard Hulbert, Irina Koretsky, Ernest Lundelius, Greg McDonald, Jerry McDonald, Charles A. Repenning, Kevin Seymour, Richard E. Tedford, and David Webb. For their helpful reviews of the manuscript or parts thereof I am grateful to the late Elaine Anderson, whom we shall all miss, to Fred Grady, Greg McDonald, Jerry McDonald, Clayton E. Ray, Charles A. Repenning, Richard E. Tedford, and David Webb. Robert E. Weems of the U.S. Geological Survey provided advice about current stratigraphic interpretations of Pleistocene deposits on the Coastal Plain of South Carolina and reviewed a portion of the manuscript. For valuable aid in the field I owe a special debt of gratitude to Peter Coleman, Albert Duc, Johnny Hanlon, Joel Padgett, Kenneth Rowland, Joseph Stephenson, Aaron Stokes, and the late Claude Newton and E.F. St. Mary. I thank Jonathan Durst-Glenn for his fine preparation of certain specimens mentioned below, and Bonnie Fagerstrom and Deborah Stokes for their assistance in proofreading the manuscript. To Clayton E. Ray I extend my deepest appreciation for many favors over the years and for his identification of several of the specimens reported in this paper. And I am sincerely grateful to Mary McDonald, Editor, American Philosophical Society, for her enduring patience and help during the production of this work.

I wish to take this opportunity to thank my photographer, Bryan Stone, for the many hours that he has contributed in making the photographs for this publication and others during more than a decade of assistance with his splendid camera work. Even the best of words could not adequately express my gratitude for his patience and his dedication to quality.

ADDITIONS TO THE PLEISTOCENE MAMMAL FAUNAS OF SOUTH CAROLINA, NORTH CAROLINA, AND GEORGIA

Albert E. Sanders [1]

INTRODUCTION

The area around Charleston, South Carolina, is one of the richest vertebrate fossil localities on the east coast of North America and was recognized as such by Louis Agassiz during his first visit to Charleston in 1847, when he saw the collection of fossils accumulated by Francis S. Holmes, a local planter with an avid interest in paleontology. At Agassiz's urging, Holmes was made curator of The Charleston Museum in 1850 and later collaborated with Michael Tuomey on the now-classic *Pleiocene Fossils of South Carolina* (Tuomey and Holmes, 1857). A pioneer in work on the paleontology of South Carolina, Holmes published the last numbers of his *Post-Pleiocene Fossils of South Carolina* in 1860, with Joseph Leidy as the author of the section on vertebrates. After the Civil War had devastated economic and scientific endeavors in South Carolina, Holmes led the way in developing the mining of phosphate in the Charleston area, which became a major factor in the recovery of the local economy. Operating from 1869 to about 1912, that industry produced large numbers of fossils and made "the Ashley River phosphate beds" one of the best known fossil localities in eastern North America.

In a special "Industrial Issue" published in 1888, the Charleston *News and Courier* gave the following account of the fossil discoveries in the phosphate beds:

> These deposits consist of nodules of phosphate of lime, thickly interspersed with the huge bones and teeth of antediluvian mammalian and marine mammoths of stupendous and gigantic proportions; the chrysonicocrisides, icthysauri, hadrosauri, zeuglodons, mastodons, stupendous giant baboons, prodigious mammoth gorillas, lizards 33½ feet long, and other graminivorous and carnivorous quadrupeds; also the squalodons, phocodons, dinotherinons, and members of the ichthaurian, saurian, and cetacean families, whales 500

[1] The Charleston Museum, 360 Meeting St.,Charleston, S.C. 29403.

feet long, sharks 200 feet long, briny leviathans, voracious marine vultures and other monster, rapacious denizens of the mighty deep—land and water animals lying in the same bed (Anonymous, 1888).

This article was published six years after Holmes's death in 1882, and one can only imagine the state of apoplexy into which it would have thrown him had he lived to read it. He had found no "stupendous giant baboons" and no "prodigious mammoth gorillas." Lost in the article's carnival-barker style of prose were the important specimens that the phosphate beds had actually produced, many of which were those of Pleistocene mammals.

Although isolated specimens of Pleistocene mammal remains from South Carolina were mentioned by Catesby (1743 [2]:vii), Drayton (1802), Harlan (1825), Tuomey (1848), Holmes, (1858), and others, the first systematic account of anything approaching a faunal assemblage was that of Leidy (1859,1860), who reported on Pleistocene vertebrate fossils from the vicinity of Charleston. Hay (1923a) provided a more complete listing of specimens from that area, and Leidy (1877), Ray (1965; 1967), Ray and Sanders (1984), Downing and White (1995), and McDonald et al. (1996) reported individual taxa. Allen (1926) gave accounts of some vertebrate fossils from the phosphate beds near Charleston, Roth and Laerm (1980) documented vertebrate fossils from Edisto Beach in Charleston County, and Bentley et al. (1995) described the Late Pleistocene Ardis local fauna from Dorchester County, South Carolina. The taxa reported from South Carolina by those authors beginning with Leidy (1860) are summarized in Table 1.

All of the previously published records of Pleistocene mammals from South Carolina have been of late Pleistocene age, the two largest assemblages being the one reported from Edisto Beach (28 taxa) by Roth and Laerm (1980) and the Ardis local fauna from Dorchester County (43 taxa) documented by Bentley et al. (1995). Within recent years, random collecting activities by Charleston Museum volunteers and excavations conducted by the writer have produced specimens of several mammals of Early and Middle Pleistocene ages. Those taxa, along with additional records from late Pleistocene units, now furnish a considerably broader picture of mammalian faunas of Irvingtonian and Rancholabrean times in the region that is present-day South Carolina.

The present paper reports new records of 37 taxa from early, middle, and late Pleistocene deposits in South Carolina, two taxa from the Early Pleistocene of North Carolina (Table 2) , and one taxon from the Late Pleistocene of Georgia. New information regarding two long-established taxa, *Neochoerus pinckneyi* (Hay, 1926) and *Alces runnymedensis* Hay, 1923, extends the temporal range of the former from late Pleistocene time back to the Mid-Pliocene and indicates that *A. runnymedensis* is referable to *Cervalces scotti*. Eight taxa are added to the 27 forms recorded in the Edisto Beach Rancholabrean fauna by Roth and Laerm (1980), and a summary of that important assemblage is given, along with some considerations of the role that the post-Wisconsin rise in sea level may have played in the great extinctions of the Rancholabrean megafauna along the Atlantic coast of the United States.

TABLE 1. Pleistocene mammal taxa reported from South Carolina by: **A**, Leidy (1860 [x^1], 1877[x^2]); **B**, Hay (1923); **C**, Allen (1926); **D**, Ray (1965 [x^1], 1967 [x^2]); **E**, Ray et al. (1968) [x^1], Ray and Sanders (1984) [x^2]; **F**, Roth and Laerm (1980); **G**, Bentley et al. (1995); **H**, McDonald et al. (1996); **I**, Downing and White (1995); **J**, this paper.

	A	B	C	D	E	F	G	H	I	J
Order Marsupialia										
Family Didelphidae										
Didelphis virginiana	X^1	X				X				
Order Insectivora										
Family Soricidae										
Blarina brevicauda							X			
Sorex cf. S. longirostris							X			
Family Talpidae										
Condylura cristata							X			
Scalopus aquaticus							X			
Order Xenarthra										
Family Dasypodidae										
Dasypus bellus						X	X			X
Family Pampatheriidae										
Holmesina septentrionalis						X	X			X
Family Undetermined										
Pachyarmatherium leiseyi									X	
Family Glyptodontidae										
Glyptotherium floridanum				X^1		X				
Family Megalonychidae										
Megalonyx jeffersonii		X				X	X			X
Family Megatheriidae										
Eremotherium sp.										X
Eremotherium laurillardi										X
(as Megatherium mirabile)		X^1	X	X						
(as cf. E. mirabile)					X					
Family Mylodontidae										
Paramylodon harlani										
(as Mylodon harlani)	X^1	X								
(as Glossotherium)						X				
Order Carnivora										
Family Mustelidae										
Mustela vison							X			
Lutra canadensis							X			
Spilogale putorius							X			
Mephitis mephitis							X			
Conepatus cf. C. robustus							X			
Family Canidae										
Canis dirus		?X				X	X			X
Urocyon cf. U. cinereoargenteus						X	X			
Family Procyonidae										
Procyon lotor	X^1	X				X	X			
Family Ursidae										
Tremarctos floridanus							X	X		X
Arctodus pristinus	X^1	X								X
Ursus americanus		X								X
Family Felidae										
cf. Smilodon fatalis		X								
Smilodon fatalis.										X
Panthera leo atrox										X
Panthera onca augusta.				X^2		X				X
Miracinonyx inexpectatus										X
(as "Felis sp. [?cougar Kerr])"			X							
Puma concolor										X
Lynx rufus		X					X			X

TABLE 1.

(cont.)	A	B	C	D	E	F	G	H	I	J
Family Odobenidae										
Odobenus rosmarus		X			X[1]			X		
(as Rosmarus obesus)	X[2]									
Odobenus cf. O. rosmarus						X				
Odobenus sp.										X
Family Phocidae										
Erignathus barbatus										X
Halichoerus grypus					X[1]	X				
Monachus tropicalis										X
Order Rodentia										
Family Sciuridae										
Spermophilus tridecemlineatus							X			
Sciurus carolinensis							X			
Glaucomys volans							X			
Family Castoridae										
Castoroides ohioensis	X[1]	X	X							
Castoroides cf. C. ohioensis						X				
Castor canadensis	X[1]	X				X	X			X
Family Cricetidae										
Oryzomys palustris							X			
Peromyscus sp.							X			
Neotoma floridana							X			
Microtus pennsylvanicus							X			
Microtus pinetorum							X			
Neofiber cf. N. diluvianus										X
Neofiber alleni							X			X
Ondatra zibethicus	X[1]	X			X					
Synaptomys cooperi							X			
Synaptomys australis							X			
Family Erethizontidae										
Erethizon dorsatum										X
Family Hydrochoeridae										
Hydrochoerus holmesi										X
Neochoerus pinckneyi					X	X				X
(as Hydrochoerus aesopi)	X[1]	X								
(as Hydrochoerus pinckneyi)		X	X							
Order Lagomorpha										
Family Leporidae										
Sylvilagus floridanus										
(as Lepus sylvaticus)	X[1]									
Sylvilagus floridanus ?	X									
Sylvilagus floridanus								X		
Sylvilagus palustris								X		
Sylvilagus sp.						X				
Order Cetacea										
Family Delphinidae										
Pseudorca crassidens										X
Tursiops cf. T. truncatus						X				
Family Physeteridae										
Physeter sp.			X							
(as Physeter vetus)										
Physeter sp.						X				
Family Balaenopteridae										
Genus et sp. indeterminate						X				
Order Perissodactyla										
Family Equidae										
Equus complicatus	X[1]	X	X							
Equus cf. E. complicatus						X				
Equus fraternus	X[1]									

TABLE 1.

(cont.)	A	B	C	D	E	F	G	H	I	J
Equus leidyi	X[1]	X	X							
Equus littoralis		X								
Equus sp.							X			X
Family Tapiridae										
Tapirus haysii	X[1]	X	X		X[2]					X
Tapirus veroensis					X[2]		X			
Tapirus cf. *T. veroensis*										X
Tapirus sp.		X				X				
Order Artiodactyla										
Family Tayassuidae										
Mylohyus nasutus							X			
(as *Dicotyles fossilis*)	X[1]									
(as *Mylohyus pennsylvanicus*)			X							
(as *Mylohyus* cf. *M. fossilis*)						X				
? *Platygonus compressus*		X								
(as *Tagassu lenis* and										
"*Tagassu* sp. indet.?")										
Family Camelidae										
Hemiauchenia cf. *H. macrocephala*										X
?*Palaeolama mirifica*										
(as *Camelops* sp.)		X								
(as *Procamelus minor*)			X							
Palaeolama cf. *P. mirifica*						X				
Palaeolama mirifica							X			
Family Cervidae										
Odocoileus virginianus	X[1]									
(as *Cervus virginianus*)										
Odocoileus virginianus		?X	X			X	X			
Rangifer cf. *R. tarandus*								X		X
Cervalces scotti										X
(as *Alces runnymedensis*)		X								
Cervus elaphus										X
(as *Cervus canadensis*)		X								
Family Bovidae										
Bison cf. *B. antiquus*						X				
Bison antiquus							X			X
Bison latifrons	X[1]	?X								
Bison cf. *B. bison*			X							
Bison sp.					X					X
Order Sirenia										
Family Trichechidae										
Trichechus sp.		X								
(as *Trichechus antiquus*)										
Trichechus sp.						X				
Order Proboscidea										
Family Mammutidae										
Mammut americanum						X	X			X
(as *Mastodon ohioticus*)	X[1]									
Mammut americanum		X								
(and as *Mammut progenium*)										
Family Gomphotheriidae										
Cuvieronius sp.										X
Family Elephanitidae										
Mammuthus columbi							X			
as *Elaphas columbi*)		X	X							
(as *Elaphas imperator*)	X									
Mammuthus cf. *M. columbi*						X				

TABLE 2. Biochronologic distribution of mammalian taxa from North Carolina (NC) and South Carolina reported in the present paper. Glacial-interglacial stages mostly follow Richmond and Fullerton(1986). Boundary of Irvingtonian/Rancholabrean Land Mammal ages is based upon *Bison* specimen from Ten Mile Hill Beds, but all other correlations of time divisions and glacial-interglacial stages are approximate. Formational abbreviations are: WAC (Waccamaw); PEN (Penholoway); LAD (Ladson); TMH (Ten Mile Hill Beds); SOC (Socastee); WANDO (Middle and Upper); UNNAMED (X, undetermined offshore unit; X[1], late Pleistocene sediments, Giant Portland Cement Company quarry, Berkeley County [source of Ardis local fauna]).

	PLEISTOCENE					
	Early			Middle	Late	
	Pre-Illinoian			Illinoian	Sangamon	Wisconsin
	Irvingtonian				Rancholabrean	
TAXON	WAC (upper bed)	PEN	LAD	TMH	WANDO (M)(U) SOC	UNNAMED
Neofiber cf. *N. diluvianus*	X (NC)					
Cuvieronius sp.	X (NC)			X	X	
Hydrochoerus holmesi	X			X	X X	
Miracinonyx inexpectatus	X	X				
Tapirus haysii	X			X		
Neochoerus pinckneyi	From Mid Blancan	- - - -→		X	X	X
Dasypus bellus		X		X		
Eremotherium sp.		X				
Equus sp.		X	X	X		
Cervus elaphus		X				X
Arctodus pristinus			X		X X?	
Megalonyx jeffersonii			X		X	X
Eremotherium laurillardi			X			X
Tapirus cf. *T. veroensis*			X			
Holmesina septentrionalis				X		X
Odobenus sp.				X		
Castor canadensis					X	
Neofiber alleni				X	X	
Bison sp.				X		X
Erignathus barbatus				X		
Monachus tropicalis				X		X
Tremarctos floridanus				X?		X
Cervalces scotti					X	
Hemiauchenia cf. *H. macrocephala*					X	
Ursus americanus					X	
Odobenus rosmarus					X	
Rangifer cf. *R. tarandus*					X	
Mammut americanum					X	
Canis dirus					X	X[1]
Bison antiquus antiquus						X
Smilodon fatalis						X
Panthera leo atrox						X
Panthera onca augusta						X
Puma concolor						X
Lynx rufus						X
Erethizon dorsatum						X
Pseudorca crassidens						X

GEOLOGIC SETTING

Sporadic studies of the Pleistocene stratigraphy of the area about Charleston have been conducted for many years, Sloan (1908), Cooke (1936), Malde (1959), and Colquhoun (1965), particularly, made significant contributions to the understanding of Pleistocene marine deposits in this region. Subsequently, exhaustive drilling surveys by Weems and Lemon (e.g., 1984a, b; 1987; 1988) produced new geologic maps that clarified a number of questions about Pleistocene formational nomenclature and depositional sequences. Although a number of stratigraphic problems remain to be resolved, their work and that of others (e.g., Bybell, 1990; McCartan et al., 1982, 1990) are currently the most accurate interpretations of the Pleistocene units on the Lower Coastal Plain near Charleston (Figure 1). Consequently, it is now possible to place reasonably reliable dates on Pleistocene vertebrate remains from that area, based primarily on coral dates obtained by Szabo (1985) and nannofossil dates by Bybell (1990) from the various deposits. Further northward along the coast to the Cape Fear region, the Pleistocene units have been mapped by Owens (1989). In that region, Pleistocene vertebrate remains have been found in the upper bed of the Waccamaw Formation and in the Socastee Formation. The following Pleistocene units have furnished fossil mammal remains discussed in the present paper:

Unnamed Units

Undetermined offshore unit, Edisto Beach, Charleston County, South Carolina
(Late Pleistocene, ca. 50,000–10,000 years before present [ybp])

For many years vertebrate fossils have been collected on the beach at Edisto Island, having been washed out of an as-yet-undetermined and unnamed stratigraphic unit deposited offshore during Wisconsinan time. Both terrestrial and marine vertebrates are found in the Edisto material, indicating the presence of two distinct faunal horizons. A U.S. Geological Survey power-auger hole put down not far from the shoreline of Edisto Beach revealed that the Pleistocene sediments unconformably overlie the early Miocene Marks Head Formation (Figure 2) but shed no light on the identity of the late Wisconsinan sediments on the now-submerged coastal plain seaward of the present beach. The deposits furnishing the terrestrial vertebrates obviously are remnants of landforms that were exposed during the periods of extremely low sea level that accompanied Wisconsinan glaciation. Since a substantial length of time would have been necessary for the buildup of topsoil on the land surfaces exposed by regression of the Sangamonian seas and for the stages of plant succession that followed the accumulation and enrichment of the topsoil, it seems unlikely that environmental conditions favorable to the terrestrial mammals represented in the Edisto material would have developed before a considerable portion of early Wisconsinan time had elapsed. Thus, we may speculate that the terrestrial mammal remains from Edisto Beach proba-

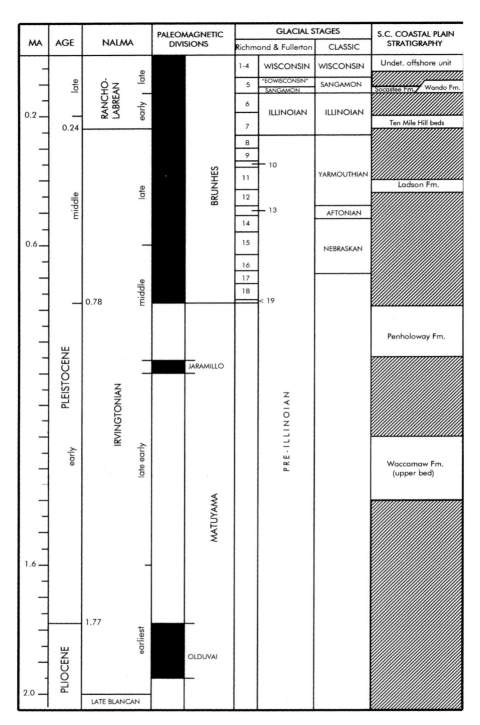

Figure 1. Correlation of Pleistocene stratigraphic units on the Coastal Plain of South Carolina (after Bybell [1990] and McCartan et al. [1990]) with Pleistocene time divisions in millions of years ago (MA) and geomagnetic polarity time scale (after Berggren et al. [1995]), glacial-interglacial stages (after Richmond and Fullerton, 1986); and North American Land Mammal Ages (NALMA). Boundary of Blancan/Irvingtonian Land Mammal Age follows Morgan and Hulbert (1995); Irvingtonian/Rancholabrean boundary is based upon *Bison* astragalus (ChM PV6865) from late middle Pleistocene reported in the present paper. Lower bed of Waccamaw Formation is not shown because of its uncertain position in the stratigraphic column.

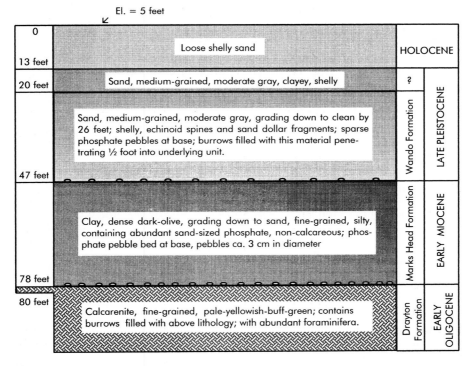

El. = 5 feet

0		
13 feet	Loose shelly sand	HOLOCENE
20 feet	Sand, medium-grained, moderate gray, clayey, shelly	?

Sand, medium-grained, moderate gray, grading down to clean by 26 feet; shelly, echinoid spines and sand dollar fragments; sparse phosphate pebbles at base; burrows filled with this material penetrating ½ foot into underlying unit.

Wando Formation — LATE PLEISTOCENE

47 feet

Clay, dense dark-olive, grading down to sand, fine-grained, silty, containing abundant sand-sized phosphate, non-calcareous; phosphate pebble bed at base, pebbles ca. 3 cm in diameter

Marks Head Formation — EARLY MIOCENE

78 feet

80 feet

Calcarenite, fine-grained, pale-yellowish-buff-green; contains burrows filled with above lithology; with abundant foraminifera.

Drayton Formation — EARLY OLIGOCENE

Figure 2. Stratigraphic sequence in 80-foot-deep U.S. Geological Survey power auger hole EI-11 at Edisto Beach State Park, Charleston County, South Carolina. Wisconsinan-age continental deposits that have furnished remains of Pleistocene vertebrates were laid down on the presently submerged glacial-age coastal plain seaward of the drill hole and thus do not appear in the log of this hole.

bly date from the Mid Wisconsinan to the Wisconsin glacial maximum. That idea is supported to some degree by an amino-acid epimerization date of 40,000 years extracted from a fragment of turtle shell (*Pseudemys*) from Edisto beach in an experimental test initiated by Robert E. Weems and I and conducted by Julia Corrado in the laboratory of Edward Hare at the Carnegie Geophysical Laboratory. Although the accuracy of amino-acid dates from bone has not been sufficiently investigated, the date cited above is certainly in keeping with the geological time frame and thus would appear to be reasonably accurate. The possibilty that the marine mammal remains could have washed out of deposits laid down on the continental shelf during the glacial maximum seems highly improbable. Consequently, they would seem most likely to date from latest Wisconsinan or early Holocene times, when sea level stands were more nearly as they are today.

Unnamed unit at Giant Cement Company quarry, Dorchester County (Late Pleistocene, ca. 19,000 ybp)

Bentley et al. (1995) reported 43 taxa of Pleistocene mammals from deposits that they referred to the Late Pleistocene on the basis of C^{14} dates of

18,940–18,530 ybp and the late Rancholabrean flavor of the assemblage, which they called the Ardis Local Fauna. The sediments suggest a swamp-like environment.

Socastee Formation (Late Pleistocene [Sangamonian], ca. 120,000 ybp [R.E. Weems, personal communication, July 2001]).

Two depostional lithofacies are present. The non-marine facies overlies the marine portion and is the source of terrestrial vertebrate remains. "The major coastal Pleistocene unit in the Cape Fear region" (Owens, 1989), it extends southward to the vicinity of Winyah Bay.

Wando Formation (Late Pleistocene, ca. 130,000–70,000 ybp).

Comprised of three unnamed members (upper, middle, and lower) deposited during Sangamonian and Eowisconsinan time. The upper beds have been dated at about 70,000 years, the middle beds at 100-85,000 years, and the lower beds at 130–120,000 years (Weems and Lemon,1984a, b; McCartan et al., 1990, R.E. Weems, personal communication, 12 July 1994). Represented by four depositional lithofacies, the Middle Wando fluvial-estuarine facies (c. 100,000 years; c. 1.5-3 m thick) being the stratum most commonly encountered in the vicinity of Charleston.

As defined by McCartan et al. (1980), the Wando Formation is the first barrier-back-barrier system landward of the modern shore and includes the sediments that were formerly considered to represent the lower part of the Pamlico Formation and terrace (Cooke, 1936; Stephenson, 1912) and the younger, more seaward "Princess Anne" and "Silver Bluff" terraces (Colquhoun, 1965; Cooke, 1936). The crests of barriers in the Wando Formation reach a maximum elevation of approximately 10 m, and the highest back-barrier deposits are approximately 5 m above modern sea level (McCartan et al., 1980).

At the bottom of the Wando there is a thick lag deposit of large phosphate nodules that was the basis of the phosphate industry that thrived in the Charleston area during the latter part of the 19th century and died out in the early 20th century. Ambiguously mentioned in the literature for many years as the "Ashley River phosphate beds," the stratigraphic unit that was the source of the land rock mined so successfully has now been determined. It is the Middle Wando, which was deposited approximately 100,000 years ago during the "Eowisconsin" time division of Richmond and Fullerton (1986:8), a period of cooling temperatures that followed the Sangamon interglacial period and "preceded widespread Wisconsin glaciation. . . . and corresponds to marine oxygen isotope substages 5d through 5a" (Figure 1).

Discovery of the unit that produced the phosphate was made during the excavation of an eremothere skeleton by the Charleston Museum in early Pleistocene sediments at a site on the old plantation called "Feteressa," a few miles northeast of Charleston on S.C. Route 642. Though it is in the center of a heavily mined area, the property on which the sloth site is located was not mined and thus furnished the first opportunity in recent times to observe the rock in place and to identify the unit from which it was obtained. The upper part of the section at the excavation site consists of a 32-inch (81.3 cm) deposit of the Wando Formation, the bottom 8 inches (20.3 mm) of which is a lag deposit of large phosphate rocks and reworked bones from the lower Wando. At this locality the Wando overlies the Penholoway Formation (Early Pleistocene) in certain areas and the Chandler Bridge Formation (Late Oligocene) in others.

In a now-filled-in borrow pit for the Mark Clark Expressway under construction in 1979, all three members of the Wando Formation were in place and unconformably overlying the Mid-Pliocene Goose Creek Limestone. The pit was located between S.C. Route 61 and Bull Creek, a short tributary of the Ashley River just north of Ashley Hall Plantation Road (Johns Island 15′ quadrangle) approximately 2.7 km (1.65 mi.) north of S.C. Route 7. This locality is in the heart of the old phosphate mining region along the Ashley River and is about 8.8 km (5.5 miles) south of Runnymede Plantation, site of the Magnolia Phosphate Mine operated by Charles C. Pinckney, Jr., who accumulated a significant collection of vertebrate fossils recovered during the phosphate mining days. That collection is discussed below and will be mentioned frequently in the following pages. The Mark Clark Pit is of particular significance because it documents the presence of all three members of the Wando at at least one locality in the phosphate mining region along Ashley River and is only a short distance south of the Magnolia Mine at Runnymede Plantation, where all three members may also be present. Thus, while most of the vertebrate fossils in the Pinckney collection may be assumed to have come from Middle Wando fluvial-estuarine facies (c. 100,000 years)—the unit most likely to have produced terrestrial vertebrates-the possiblity that some were reworked from the lower member must also be considered.

Ten Mile Hill Beds (Late Middle Pleistocene, ca. 240,000–200,000 ybp)

Four depositional lithofacies. The 5-10-m-thick fluvial-estuarine clayey sand and clay facies is the stratum that produces terrestrial vertebrate remains.

Ladson Formation (Middle Pleistocene, ca. 450,000–400,000 ybp)

Three depositional lithofacies. Terrestrial vertebrate remains are found in the 3–6-m-thick fluvial-estuarine facies.

Penholoway Formation (Early Pleistocene, ca. 925,000–700,000 ka [Bybell, 1990:B7])

Three depositional lithofacies. Terrestrial vertebrate remains have been found in the fluvial-estuarine facies and marine vertebrates are known from the marine facies.

Waccamaw Formation (Upper bed) (Early Pleistocene, 1.37–1.2 Ma)

There are contrasting opinions concerning the age of the upper bed of the Waccamaw Formation. Owens (1989) placed the Waccamaw in the Lower Pleistocene, following Blackwelder's (1981) belief that this unit was deposited during a subtropical interval of the Pleistocene. Cronin (1990:C4) placed the Waccamaw in the Early Pleistocene along with the Wicomico, Penholoway, and James City formations, ranging them from 2.0 to 0.7 Ma. Commenting on Stanley's (1986:20) assignment of the Waccamaw and the Caloosahatchee Formation of Florida to the late Pliocene based on his judgement that a regional mass extinction beginning in the Late Pliocene had eliminated about 65 percent of Pliocene forms by early Pleistocene time, Cronin (1990:C19) stated that "it is known from planktonic foraminifers and nannofossils (Akers, 1972; Cronin and others, 1984) that the Waccamaw and Caloosahatchee Formations are early Pleistocene in age, no older than the Olduvai event 1.8 to 1.6 Ma, near the Pliocene-Pleistocene boundary." He also took issue with Stanley's (1982:186) assertion that "both faunas lived before major glacial episodes and accompanying sea level depressions," claiming that "The Waccamaw and Caloosahatchee molluscan faunas do not predate the first hemispheric glaciation" (Cronin, 1990:C19). Bybell (1990:B7) examined the nannoplankton in a sample from the upper bed approximately 5.8 km southwest of Calabash, and concluded that the absence of *Cyclococolithus macintyrei* and the presence of *Helicosphaera sellii* "presumably places this sample within [nannoplankton] Zone NN19b or about 1.37 to 1.2 Ma." Her best evidence for an early Pleistocene age for the upper Waccamaw unit appears to be the presence of *Gephyrocapsa oceanica*, which her chart of nannoplankton stratigraphic ranges (Bybell, 1990:B6, fig. 3) does not show occurring in beds below the Pliocene-Pleistocene boundary.

New evidence presented in the *Neofiber* cf. *N. diluvianus* account below strongly supports an early Pleistocene age for the upper bed of the Waccamaw formation .

PRINCIPAL PLEISTOCENE MAMMAL ASSEMBLAGES
FROM SOUTH CAROLINA

C.C. Pinckney, Jr., Collection

In 1957 The Charleston Museum received a collection of 198 vertebrate fossils (ChM PV2506-PV2704) accumulated by Charles C. Pinckney, Jr., owner of the Magnolia Phosphate Mine, which operated from 1869 (Anon., 1873:60) until its closing in 1910 (personal communication, Sally Pinckney Burton, 1976). O.P. Hay examined this collection during his visit to Charleston in 1915, and the recognizable Quaternary mammal taxa represented in it were included in his list of Pleistocene mammals from South Carolina (Hay, 1923a). At that time, the Pleistocene stratigraphic units which are their probable origin were virtually unknown, but it is now possible to clarify that matter considerably.

Located on the Runnymede Plantation property on Ashley River Road (S.C. Route 61) some 11 miles northeast of Charleston, the mine was of the open-pit variety, consisting of long parallel trenches dug by hand into the Wando Formation, at the bottom of which there is a lag deposit of large phosphate nodules and re-worked bones and teeth. Although Pinckney did not record the precise localities for most of the specimens in his collection, the majority of them were found during the phosphate mining days, so there is good reason to believe that most, if not all, of them were found during excavations at the Magnolia mine or in dredging for phosphate nodules in the nearby Ashley River. The original stratigraphic origin of many of the specimens is not always clear, although many seem almost certainly to have come from the lag deposit at the base of the Wando Formation because there is no alternative Pleistocene stratigraphic source present in the area at the old Magnolia mine. There are four Pleistocene units within a ten-mile radius of Runnymede Plantation, *viz.*, the early Pleistocene Penholoway Formation, the Ladson Formation and the Ten Mile Hill Beds, both of middle Pleistocene age, and the late Pleistocene Wando Formation. Fossil bones found in place in the Penholoway Formation are medum to dark brown in color, while those from the Ladson and the Ten Mile Hill Beds are usually of a light brown or buff color, sometimes with light orange or black iron stains. As noted in the *Megalonyx* account in the present paper, the only specimens yet found in place in the Wando Formation are light gray and are very light in weight, most of the calcium having been leached from them. Most specimens from the Wando Formation are well mineralized and almost uniformly black, particularly those from the lag deposit at the base. Since virtually all of the specimens in the Pinckney collection are very dark grey or black, they seem most likely to have come from the basal lag deposit of the Wando. The Pleistocene mammal specimens from the lag deposit represent Rancholabrean taxa almost exclusively and thus were probably reworked from basal Wando sediments no longer present. There is certainly a possibility that some of the bones in the lag deposit were reworked from the Ladson Formation or the Ten Mile Hill Beds and thus may be of Irvingtonian age, but such specimens would be almost impossible to reognize and to verify. One notable exception is the holotype molar of *Neochoerus pinckneyi* (Hay, 1923) (ChM PV2506), which was in Pinckney's possesion at the time of its original description by Hay (1923) as *Hydrochoerus pinckeyi*. As discussed in the *Neochoerus* account in the present paper, sediment adhering to that specimen demonstrate that it is much older than the Pleistocene age previously accorded it by Hay (1923). But other adherent material indicates that it was dredged from older beds in the Ashley River and thus would not be typical of the specimens recovered during the land-mining operations. Since the vast majority of the Pinckney specimens seem clearly to have been found during the land excavations into the Wando Formation, this material is regarded in the present paper as being of Wando age. The Pinckney collection also furnished the type specimen of *Alces runnymedensis* Hay, 1923 (ChM PV2551), which is referred to *Cervalces scotti* (Lydekker, 1898) in this paper.

Edisto Rancholabrean Fauna

Fossil remains of Pleistocene vertebrates, mostly those of mammals, have been found along the beach on Edisto Island (USGS Edisto Beach 7.5′ quadrangle), a barrier island southwest of Charleston, for many years. Roth and Laerm (1980) reported specimens representing 27 mammalian taxa (23 terrestrial forms and four marine mammals) from that locality in the Charleston Museum collection. Eight additional taxa (six terrestrial forms, one cetacean, and one pinniped) from Edisto Beach are documented in the present paper, bringing the total to 35 taxa known from this locality. The presence of marine mammals (three cetaceans and three pinnipeds) among the Edisto specimens attests to the mixed nature of the material and mitigates against the designation of the Edisto mammal assemblage as a local fauna. However, the terrestrial mammals are forms characteristically found in Rancholabrean local faunas elsewhere in the eastern United States (e.g., Florida), so it is proposed here that this portion of the taxonomic assemblage from Edisto Beach be known informally as the Edisto Rancholabrean fauna. The significance of the Edisto vertebrate material is discussed in the Summary and Conclusions.

Ardis Local Fauna

Bentley et al. (1995) reported 43 mammalian taxa from Late Pleistocene sediments filling solution pits in the top of the late Eocene Harleyville Formation and in an overlying stratum of sand in the Giant Cement quarry near Harleyville in Dorchester County South Carolina. This fauna is of particular interest because of the assemblage of smaller mammals that it contains (e.g., *Blarina, Sorex, Scalopus, Condylura, Spilogale, Mephitis, Glaucomys, Sciurus,* and two species of *Microtus*) as well as for the presence of certain forms now occurring only in other regions of the United States (*Conepatus, Spermophilus,* and two species of *Synaptomys*). The age of the fauna is late Pleistocene (late Rancholabrean) based upon C^{14} dates of 18,940–18,530 (maximum of ca. 22,000) ybp obtained from mammal and reptile bone apatite (Bentley et al., 1995:4).

MATERIALS AND METHODS

With some exceptions based on more recent interpretations (e.g., McKenna and Bell, 1997), the nomenclature of mammalian taxa discussed in the present paper follows Kurtén and Anderson (1980). So that all of the marine mammal localities could be placed on one map, I have taken the account of the odontocete *Pseudorca crassidens* out of systematic order (preceding Artiodactyla) and have placed it after the pinnniped accounts. Otherwise, the systematic arrangement follows that of Kurtén and Anderson (1980). In the main I have followed the interpretations of Pleistocene glacial and interglacial stages proposed by Richmond and Fullerton (1986), but for the sake of diagrammatic convenience I have omitted their "Eowisconsinan" stage from Table 2. The nomenclature and estimated ages of geologic formations mentioned in the present paper follow Bybell (1990), McCartan et al. (1990), Owens (1989), Weems and Lemon (1984a, b; 1987; 1988, 1989), and Weems et al. (1997). Measurements of specimens examined were taken with a standard dial caliper.

Maxillary dentition is indicated by upper case lettering (M3) and mandibular dentition by lower case (m3).

To facilitate discussion and future recognition of specimens borrowed from private collections, a number prefixed by the owner's initials was placed on each specimen with the permission of the owner.

Some workers may not agree with the utilization of specimens in private collections in a scientific publication; in fact, one of the reviewers of this paper has expressed concern about the privately owned material included in it. He has observed quite correctly that there is no guarantee that such material will be available to the scientific community for future study.

Many paleontologists can cite specific examples of important specimens in private collections that have been lost or sold, never to be seen again. Without question, the use of privately-owned fossils for scientific purposes is generally a most unwise practice; however, there may be exceptional cases in which one should report significant new information that might well be lost if no effort is made to record it. In the writer's opinion, the felid specimens from Edisto Beach reported in the present paper exemplify just such a situation. Previously, only one felid taxon, *Panthera onca augusta,* had been recorded from Edisto Beach (Ray, 1965). This paper reports four additional taxa from that locality—*Smilodon fatalis, Panthera leo atrox, Puma concolor,* and *Lynx rufus.* The latter form, and *Panthera onca augusta,* are represented by Charleston Museum specimens, but the other three taxa are documented by specimens in three private collections. Thus, of the five felids now known to have occurred in the Pleistocene fauna of the Edisto Beach region, three (60%) are represented by privately-owned material. If those specimens were to be simply ignored, the omission of such a large percentage of the felids represented in the fossil remains from Edisto Beach would conceal important distribution records for felid taxa in the southeastern United States and would seriously compromise any appraisals of predator-prey relationships among the Pleistocene mammals of the Edisto paleoenvironment. For those reasons, the privately-owned specimens have been included, albeit with considerable reluctance. Another privately-owned specimen, a mandible of *Erithizon dorsatum*, is reported because it is the first record of that taxon from South Carolina. Sadly, that specimen and others reported from the private collection of Don Marvin of Edisto Beach may already be in danger of loss to the scientific community. Mr. Marvin's recent death has left his collection uncommitted to an institutional collection, and it seems destined to be distributed among family members. So we have still another case of the instability of private collections and the probable loss of important material therein. Perhaps there is some comfort in the fact that the most important material in that collection has been recorded in the present paper and is documented by casts and/or photographs.

Institutional acronyms and initials of private collectors lending material to this study are as follows:

AMNH—American Museum of Natural History; **ANSP**—Academy of Natural Sciences of Philadelphia; **ChM**—The Charleston Museum; **ISM**—Illinois State Musem; **MCZ**—Museum of Comparative Zoology, Harvard University; **ROM**—Royal Ontario Museum; **SCSM**—South Carolina State Museum; **UF**—Florida Museum of Natural History; **USNM**—U.S. National Museum of Natural History; **RBH**—Ray and Bettie Harkless; **LH**—Lee Hudson; **CK**—Chet Kirby; **TL**—Terry Lee; **DM**—Don Marvin; **NR**—Ned Riddle.

Abbreviations: **Ma**—milllions of years ago; **ka**—thousands of years ago; **ybp**—years before present; **S.C.**—South Carolina; **Co.**—County; **Fm.**—Formation

SYSTEMATIC PALEONTOLOGY

Class MAMMALIA
Infraclass EUTHERIA
Superorder XENARTHRA
Order CINGULATA
Superfamily DASYPODOIDEA
Family DASYPODIDAE
Subfamily DASYPODINAE
Genus *Dasypus* Linnaeus, 1758
DASYPUS BELLUS (SIMPSON, 1929)
Figures 3, 4a-c

MATERIAL.—ChM PV5807, PV5825, moveable osteoderms; ChM PV5838, immoveable osteoderm.

LOCALITY AND HORIZON.—ChM PV5807, PV5825: S.C., Charleston Co.; in association with megathere remains excavated on east bank of lake at Trailwood Trailer Park, east side of S.C. Route 642 (Dorchester Road), c. 12 km northwest of Charleston; A.E. Sanders et al., November 1982; Penholoway Fm., Early Pleistocene. ChM PV5838: S.C., Dorchester Co.; ditch in Irongate subdivision, c. 1.7 mi. (2.73 km) northeast of Dorchester Road (S.C. Route 642); Vance McCollum, summer 1980; Ten Mile Hill Beds, Late Middle Pleistocene.

AGE.—ChM PV5807, PV5825, Irvingtonian, Pre-Illinoian; ChM PV5838, Early Rancholabrean, Late Illinoian.

DISCUSSION.—This extinct relative of the modern armadillo, *Dasypus novemcinctus*, has been recorded from the Late Pleistocene of South Carolina by Roth and Laerm (1980:11) and by Bentley et al. (1995:6). Two moveable osteoderms (ChM PV5807, PV5825, Figures 4a-b), found during the excavation of a partial skeleton of *Eremotherium* by a Charleston Museum party, are the first early Pleistocene records of *D. bellus* in the state. PV5807 is 20.5 mm in length and 9.3 mm in width posteriorly. PV5825 is 32.1 mm long and 13.1 mm in greatest width.

An immoveable osteoderm (ChM PV5838), collected by Vance McCollum from the Ten Mile Hill Beds in a ditch bank in Dorchester County, approximately 22 mi. (37 km) northwest of Charleston, documents *D. bellus* in the late middle Pleistocene of South Carolina.

D. bellus ranged from Blancan through Rancholabrean time and is known from many sites in North and South America (Kurtén and Anderson, 1980:130–131).

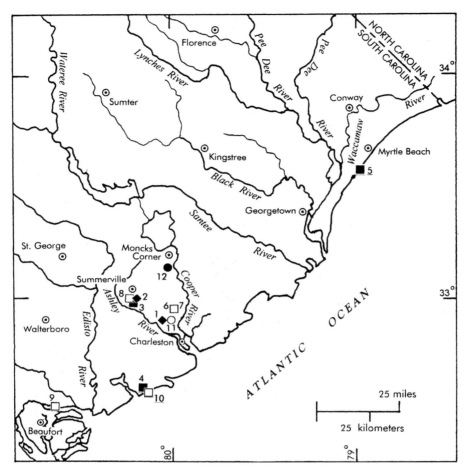

Figure 3. Locality records in South Carolina for *Daysypus bellus* (◆ 1-2); *Holmesina septentrionalis* (■ 3-5); *Meagalonyx jefforsonii* (□ 6-10); *Eremotherium* sp. (○ 11); and *Eremotherium laurillardi* (● 12) reported in the present paper.   **1**, ChM PV5807, PV5825, Charleston Co., Trailwood Trailer Park, Penholoway Fm. (Early Pleistocene); **2**, ChM PV5838, Dorchester Co., Irongate subdivision, Ten Mile Hill Beds (Late Middle Pleistocene). **3**, ChM PV5837, Dorchester Co., Trolley Road, Ten Mile Hill Beds (Late Middle Pleistocene); **4**, ChM PV2032, PV2416, PV2420, PV2785-2786, PV3458, PV4789, PV4962-4979, PV5005, PV5669, PV5851, Charleston Co., Edisto Beach, undetermined offshore unit (Late Pleistocene); **5**, ChM PV4880, Horry Co., Garden City Beach, undetermined offshore unit (Late Pleistocene). **6**, ChM PV4930, Charleston Co., Goose Creek, Ladson Fm. (Middle Pleistocene); **7**, ChM PV5847, Berkeley Co., Goose Creek, Ladson Fm. (Middle Pleistocene); **8**, ChM PV5853, Dorchester Co., Irongate subdivision, Wando Fm. (Late Pleistocene); **9**, ChM PV5850, Beaufort Co., Coosaw River, Wando Fm.? (Late Pleistocene); **10**, ChM PV5706, PV5708, Charleston Co., Edisto Beach, undetermined offshore unit (Late Pleistocene). **11**, ChM PV4748, Charleston Co., Trailwood Trailer Park, Penholoway Fm. (Early Pleistocene). **12**, ChM PV4803, Berkeley Co., near Moncks Corner; Ladson Fm. (Middle Pleistocene).

Family PAMPATHERIIDAE
Genus *Holmesina* Simpson, 1930
HOLMESINA SEPTENTRIONALIS (LEIDY, 1889)
Figures 3, 4d-e, 5

MATERIAL.—ChM PV2032, PV2416, PV2785-PV2786, carapacial osteoderms (Roth and Laerm, 1980:12), PV 2419, carapacial osteoderm (Roth and Laerm, 1980;12, as "PV2619;" also a non-existent "PV2620,"), PV2420, caudal ring scute (Roth and Laerm, 1980); PV5841, right mandibular ramus; ChM PV5669, portion of left mandibular ramus; PV4879, PV4880, PV4962-PV4979, PV5005, PV5837, carapacial osteoderms; PV3458, PV5851, moveable osteoderms.

LOCALITY AND HORIZON.—Chm PV5837: S.C., Dorchester Co.; ditch behind shopping center on north side of Trolley Road (County Road 199), 0.2 mi. (0.3 km) east of S.C Route 642; Vance McCollum, 13 December 1981; Ten Mile Hill Beds, Late Middle Pleistocene. ChM PV4880: S.C., Horry Co.; Garden City Beach; Doris Holt , 7 June 1992; undetermined offshore unit, Late Pleistocene. S.C., Charleston Co.; Edisto Beach, undetermined offshore unit, Late Pleistocene: ChM PV5841, H.M. Rutledge, summer 1936; ChM PV5669, PV2032, PV2416, E.R.Cuthbert, Jr., c. 1960 (PV5669) and summer 1969 (PV2032, PV2416); ChM PV2419, the Misses Holmes, c. 1940. ChM PV3458, Margaret Weeks, c. 1980. ChM PV2420, PV2785-2786, PV4879, PV4962-PV4979, Doris Holt, various dates; ChM PV5005, Shirley and Leonard Pazucha, c. 1979; ChM PV5851, J.H. Garety, 19 June 1935.

AGE.—ChM PV5837, Early Rancholabrean, Late Illinoian; ChM PV5841, PV2416, PV2419, PV2420, PV2785-2786, PV4879, PV4962-PV4979, PV3458, PV4879, PV4880, PV4962-PV4979, PV5005, PV5669, PV5851, Rancholabrean, Wisconsinan.

DISCUSSION.—The only pampathere known to have occurred in the southeastern United States during the Late Pleistocene (Edmund, 1995), *H. septentrionalis* has been reported from South Carolina by Roth and Laerm (1980:12), based on carapacial scutes in The Charleston Museum, and by Bentley et al. (1995:6–7), who recorded two osteoderms in the Ardis local fauna. Unfortunately, some of the ChM catalogue numbers were cited erroneously by Roth and Laerm (1980); hence, all of the specimens that they reported are included above with their correct numbers.

ChM PV5841, a right mandibular ramus (Figure 5a, c), was on loan at the time that Roth and Laerm (1980) were studying the Charleston Museum specimens from Edisto Beach and thus was not included in their report on the Edisto assemblage. Collected on Edisto Beach by H.M. Rutledge in 1936, this specimen is missing its anterior end beyond the alveolus for the 4th tooth and lacks both the corocoid and condylar regions. The alveolae for the 4th through the 9th teeth are preserved, and the bases of the 5th, 6th, 7th, and 8th teeth are present, the crowns having been broken off. As preserved, the specimen is 180.4 mm in anteroposterior length.

A less complete portion of a left mandibular ramus (ChM PV5669, Figures 5b-c) was collected on Edisto Beach by Edmund R. Cuthbert, Jr., sometime during the 1960s. Lacking the anterior end beyond the 6th tooth and the posterior region behind the anterior root of the 9th tooth, this specimen preserves portions of the 6th,

Figure 4. Scutes of *Dasypus bellus* (**A, B, C**) and *Holmesina septentrionalis* (**D, E**) from South Carolina. **A**, ChM PV5807, **B**, ChM PV5825, moveable osteoderms, Charleston Co., Penholoway Fm. (Early Pleistocene); **C**, PV5838, immoveable osteoderm, Dorchester Co, Ten Mile Hill Beds (Late Middle Pleistocene); **D**, ChM PV5837, carapacial osteoderm, Dorchester Co., Ten Mile Hill Beds (Late Middle Pleistocene); **E**, ChM PV2420, caudal ring scute, Charleston Co., undetermined offshore unit (Late Pleistocene). Scale bar = 10 mm.

7th, and 8th teeth and is 101.8 mm in anteroposterior length. It seems to represent a slightly larger individual than the preceding specimen, the depth of the ramus at the 7th tooth being 55.2 mm compared to 51.2 mm in PV5841. These two rami appear to be the only skeletal elements of pampatheres yet found in South Carolina.

Additional records of *H. septentrionalis* from Edisto Beach consist of two moveable osteoderms found by J.H. Garety (ChM PV5851) and Margaret Weeks (ChM PV3458), an osteoderm (ChM PV5005) found by Shirley and Leonard Pazucha, and 19 osteoderms (ChM PV4879, PV4962-4979) collected by Doris Holt, who also found one (ChM PV4880) on the beach at Garden City, Horry County, South Car-

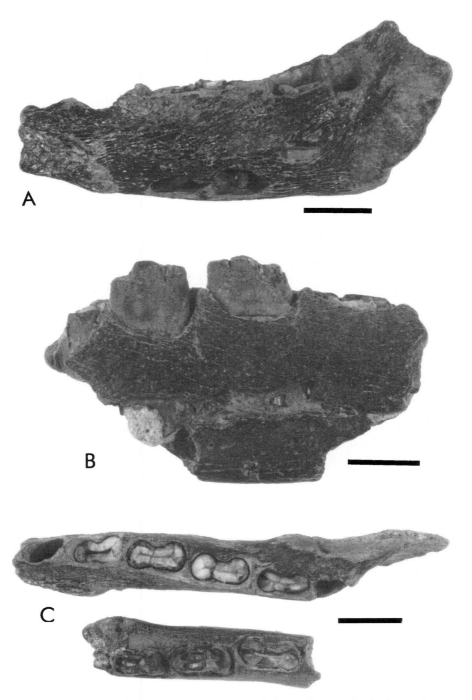

Figure 5. Mandibular rami of *Holmesina septentrionalis* from Edisto Beach, Charleston County, S.C.; undetermined offshore unit (Late Pleistocene). **A**, ChM PV5841, right ramus, lingual view; scale bar = 30 mm. **B**, ChM PV5669, portion of left ramus, lingual view; scale bar = 20 mm. **C**, ChM PV5841 (above) and ChM PV5669 (below), occlusal views; scale bar = 30 mm.

olina. A single osteoderm (ChM PV5837) from the Ten Mile Hill Beds in a ditch on Trolley Road (County Road 199) in Dorchester County is the first record of this species in the Middle Pleistocene of South Carolina.

* * *

Order PILOSA
Family MEGALONYCHIDAE
Subfamily MEGALONYCHINAE
Genus *Megalonyx* Jefferson, 1799
MEGALONYX JEFFERSONII (DESMAREST, 1822)
Figures 3, 6, 7

MATERIAL.—ChM PV4930, upper right second molariform tooth; ChM PV5847, lower left third molariform; ChM PV 5706, PV5850, upper left caniniform tooth; PV5708, upper right canininiform; ChM PV5853, four articulated thoracic vertebrae with rib.

LOCALITY AND HORIZON.—ChM PV4930: S.C., Charleston Co.; drainage ditch on south side of County Road. 996, 1.3 km southwest of U.S. Route 176 in Goose Creek; Jonathan Geisler, Bricky Way, 21 Nov.1992; Ladson Formation, Middle Pleistocene. ChM PV5847: S.C., Berkeley Co.; west side of U.S. Route 176, 0.8 km north of U.S. Route 52 in Goose Creek; Bill Palmer, 20 July 1997; Ladson Formation, Middle Pleistocene. ChM PV 5706, PV5708: S.C., Charleston Co.; Edisto Beach; E.R. Cuthbert, Jr., 1960s; undetermined offshore unit, Late Pleistocene. ChM PV5850: S.C., Beaufort Co.; Coosaw River; Earl Sloan, c. 1905; undetermined unit, Late Pleistocene. ChM PV5853: S.C., Dorchester County; ditch in Irongate subdivision west of Trolley Road (County Road 199), c. 0.8 km north of Dorchester Road (S.C. Route 642); A.E. Sanders, summer 1981; Wando Fm., Late Pleistocene.

AGE.—ChM PV4930, PV5847, Irvingtonian, Pre-Illinoian; ChM PV5853, Rancholabrean, Eowisconsinan; ChM PV5850, Rancholabrean, Sangamonian?; ChM PV 5706, PV5708, Rancholabrean, Wisconsinan.

DISCUSSION.—This moderate-sized ground sloth is known from many Pleistocene localities in the United States and is the only sloth that ranged into Canada and Alaska (Kurtén and Anderson,1980; McDonald et al., 2000). *M. jeffersonii* was first reported from South Carolina by Hay (1923a:363) from "the region about Beaufort," South Carolina, on the basis of an upper left caniniform tooth (ChM PV5850) recovered during dredging operations in the Coosaw River, as recorded by Hay on page 15 of the notes of his visit to Charleston in 1915. Roth and Laerm (1980:12) reported a mandibular fragment (ChM PV2421), four teeth (PV2012, PV2423, PV2455, PV2743), an ungual phalanx (PV2424), and a cast of an ungual phalanx (PV2428) of "*Magolonyx* [sic] cf. *M. jeffersonii*" in the Charleston Museum, all from Edisto Island. Subsequently, Bentley et al. (1995:6) recorded a tooth of *M. jeffersonii* from the Late Pleistocene Ardis local fauna in Berkeley County.

Two additional specimens from Edisto Island are available. One is an upper left caniniform (ChM PV5706) and the other is an upper right caniniform (ChM PV5708), both collected by Edmund R. Cuthbert, Jr., sometime during the 1960s.

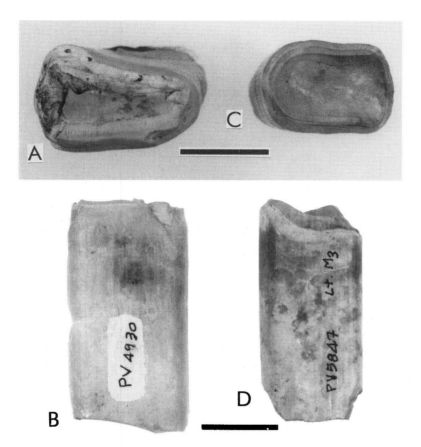

Figure 6. Teeth of *Megalonyx jeffersonii* from Ladson Formation (Middle Pleistocene) of South Carolina. **A**, ChM PV4930, upper right second molariform tooth, occlusal view; **B**, anterior view; Charleston Co., Goose Creek.. **C**, ChM PV5847, lower left third molariform tooth, occlusal view; **D**, anterior view; Berkeley Co., Goose Creek.. Scale bars = 20 mm.

Two molariform teeth from the Middle Pleistocene Ladson Formation near the small community of Goose Creek in Berkeley County provide the first Irvingtonian records of *M. jeffersonii* in South Carolina. ChM PV4930, a right upper second molariform (Figures 6a-b), was collected by Jonathan Geisler and Bricky Way in a deep drainage ditch on the south side of County Road 996, 1.3 km southwest of U.S. Route 176 . A left lower third molariform (ChM PV5847, Figures 6c-d) was found by Billy Palmer in a ditch on the west side of U.S. Route 176 approximately 0.8 km north of U.S. Route 52.

Four articulated thoracic vertebrae and two fragments of one of the acetabular regions of the pelvis of *M. jeffersonii* (ChM PV5853; Figure 7) from the late Pleistocene Wando Formation are the first known specimens of this taxon found in place in South Carolina. Each of the two anteriormost vertebrae has the neural arch fairly well intact but lacks the spinous process. The third vertebra has only a portion of the neural arch preserved, and the fourth one is represented only by a portion of the

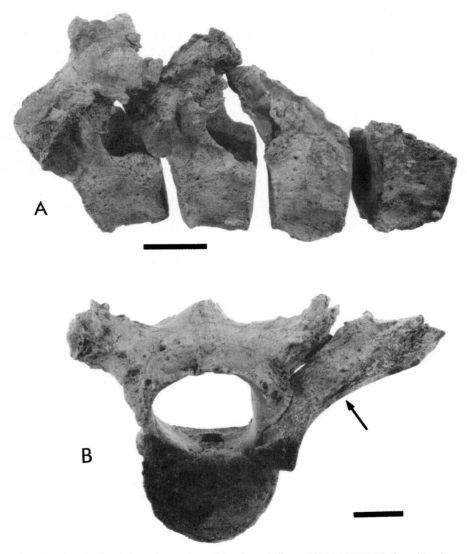

Figure 7. A, articulated thoracic vertebrae, *Megalonyx jeffersonii* (ChM PV5853), found in place in Wando Formation (Late Pleistocene), Berkeley Co. **B**, anterior view of anteriormost specimen in group; arrow points to rib head positioned as found in place. Scale bars = 50 mm.

centrum. Based on the relatively large size of the neural canal and the position of the transverse process laterally instead of ventrally on the anteriormost vertebra, this group of vertebrae appears to have come from the posterior portion of the thoracic series (Gregory McDonald, personal communication, April 2002). There is virtually no mineralization of these bones, and a significant amount of calcium has been leached from them, making them exceptionally light in weight. They are whitish-grey in color, contrasting greatly to the well-mineralized dark grey or black specimens found in the lag deposit at the base of the Wando Formation. These specimens are of considerable importance because they provide the first comparison that we have had between bone preserved in place in the Wando Formation and the preservation of bones found in the lag deposit. The age of the Wando at this locality is approximately 100,000 years (Weems and Lemon,1984a), placing it in Middle Rancholabrean time and in the Eowisconsin glacial interval of Richmond and Fullerton (1986).

Kurtén and Anderson (1980:136–137) recognized three species in the genus *Megalonyx*: *M. leptostomus* of the Blancan and early Irvingtonian, the smaller Irvingtonian form *M. wheatleyi*, and *M. jeffersonii* of the Rancholabrean. Although the two teeth from the Ladson Formation are of late Irvingtonian age there are no apparent differences between them and comparable elements of the dentition of *M. jeffersonii* as figured by Leidy (1855: plates 3, 5, 6); thus, they have been referred to that taxon. If those determinations are correct, *M. jeffersonii* was a constituent of both the Late Irvingtonian and Rancholabrean faunas of South Carolina.

* * *

Family MEGATHERIIDAE
Subfamily MEGATHERIINAE
Genus *Eremotherium* Spillman, 1948
EREMOTHERIUM SP.
Figures 3, 8

MATERIAL.—ChM PV4748, partial skull, mandibles, and elements of postcranial skeleton.

LOCALITY AND HORIZON.—S.C., Charleston Co.; east bank of lake at Trailwood Trailer Park, 183 m. southeast of Ree Street, east side of South Carolina Route 642 (Dorchester Road), c. 12 km northwest of Charleston; A.E. Sanders, P.S. Coleman et al., November 1982; Penholoway Formation, Early Pleistocene.

AGE.—Irvingtonian, Pre-Illinoian.

DISCUSSION.—In the fall of 1982, the late Claude Newton, maintenance superintendant for Trailwood Trailer Park near Charleston and a Charleston Museum volunteer preparator, found the humerus of a giant ground sloth in the bank of a newly dug lake in the trailer park, some 12 km northeast of Charleston. During the removal of that specimen by a Charleston Museum party, other bones were encountered, and permission to excavate all of the remains was kindly granted by Douglas Truluck, owner of Trailwood Park.

The upper part of the section at the excavation site consists of a 32-inch (81.3 cm) deposit of the Late Pleistocene Wando Formation, the bottom 8 inches (20.3 mm) of which was a lag deposit of large phosphate rocks and reworked bones. This deposit is the zone that was mined so extensively along the Ashley River during the latter portion of the 19th century and into the early 20th century. Immediately below it is a 38–46-cm layer of the Penholoway Formation of Early Pleistocene age, and it was in the deeper part of this bed that the sloth remains were found. In the upper part of this deposit, intermittent shell beds approxmately 15 cm thick yielded a rich assemblage of marine mollusks (ChM PI4311-PI6088). Abundant specimens of the bivalve *Carditamera arata*, essentially a Waccamaw form that did not survive beyond the Early Pleistocene, demonstrate an early Pleistocene age for the sloth-bearing bed. Certain mollusks usually found in the upper bed of the Waccamaw are absent in the Trailwood Park assemblage, indicating that it is younger than the Waccamaw and thus correlates well as a Penholoway fauna. Calcareous nannoplankton samples and a coral date have been combined to provide an age of 925,000 to 700,000 years for the Penholoway Formation (Bybell, 1990:B7). Uncomfortably underlying the Penholoway is the Ashley Formation, a Late Oligocene marine unit occurring throughout the Charleston area. At the base of the Penholoway at this site there is a lag deposit consisting of lumps of the Ashley Formation.

Completed during the month of November 1982, the excavation yielded the posterior half of the skull (Figure 8), both mandibles, five teeth, the right humerus, right and

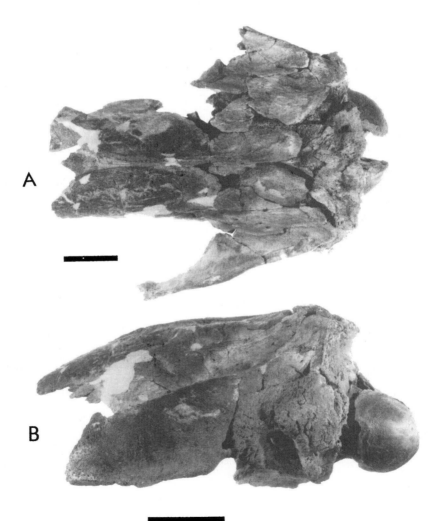

Figure 8. ChM PV4748, partial skull, *Eremotherium* sp. **A**, dorsal view; scale bar = 60 mm. **B**, left lateral view; scale bar = 50 mm. S.C., Charleston Co., Penholoway Fm.(Early Pleistocene).

left radii, the right femur, right tibia and fibula, right astragulus, right and left 5th metatarsals, eight vertebrae, elements of 16 ribs, and the pelvis, the latter in poor condition. They are the remains of a relatively small individual, compared to the size attained by the largest of the eremotheres, which "may be compared in size with a mammoth or mastodon" (Gazin, 1957:348). This specimen is in need of independent study in much greater detail because certain aspects of the skull suggest that it is not referable to *Eremotherium laurillardi*. It may represent an undescribed form or is perhaps assignable to the new species of *Eremotherium* found in the Leisey Shell Pits and other early Pleistocene deposits in Florida (Morgan and Hulbert, 1995) and recently described by De Iuliis and Cartelle (1999) as *Eremotherium eomigrans*. These discoveries are of particular significance because they are the earliest evidence of megatheres in North America.

Remains of four other mammalian taxa were found in the Penholoway deposits during the excavation; *viz.*, an osteoderm of *Dasypus bellus*, five specimens of *Equus* sp., and a vertebra of a cervid. These specimens are discussed elsewhere in the present paper. Fragments of the shells of four or five species of freshwater turtles also were found. The matrix in which the vertebrate remains were found is composed largely of coarse subangular quartz grains typical of a fluvial deposit, documenting the presence of a rather large stream that flowed through this area during Penholoway time and accounting for the presence of the freshwater turtle remains. Most of the sloth bones were found in a depression, suggesting that the animal may have become mired in deep mud at the edge of the stream, its large elongate foot perhaps preventing it from extricating itself before expiring from starvation and/or the actions of predators. The mammal and turtle remains from this site are the first Early Pleistocene vertebrate fossils reported from South Carolina.

* * *

EREMOTHERIUM LAURILLARDI (LUND, 1842)
Figures 3, 9

MATERIAL.—ChM PV4803, partial postcranial skeleton; ChM PV5842, fragment of tooth.

LOCALITY AND HORIZON.—ChM PV4803: S.C., Berkeley Co.; 68 m NW of U.S. Route 52, 3.86 km SW of old U.S. Route 52 in Moncks Corner (33° 10.1' N., 81° 01.7' W., U.S.G.S. Moncks Corner 7.5′ quadrangle); A.E. Sanders and party, May-June 1975; Ladson Formation, Middle Pleistocene. ChM PV5842: S.C., Dorchester Co.; ditch opposite O.K. Tire Store, Trolley Road (County Road 199), 0.2 km E. of Dorchester Road (S.C. Route 642; Ten Mile Hill Beds, Middle Pleistocene.

AGE.—Late Irvingtonian, Pre-Illinoian.

DISCUSSION.—The first definite report of giant ground sloth remains in South Carolina appears to be F.S. Holmes' (1848:663) brief mention of "*Megatherium*" as one of the forms represented among the fossils found in the Charleston area. Subsequently, Leidy (1855:49–55), mentioned two small fragments of teeth "from the shores of Ashley River," Charleston County, along with his discussion of "*Megatherium*" remains from Skiddaway Island, Georgia, for which he erected the new species *M. mirabile* (Leidy, 1855:17). Nearly 100 years later, Paula Couto (1954) referred *M. mirabile* to the genus *Eremotherium* Spillman, 1948, described from Pleistocene deposits on the peninsula of Santa Elena, Ecuador. As noted by Gazin, (1957:350), the manus of *Megatherium* has four digits while there are only three in the manus of *Eremotherium.*

The systematics within the genus *Eremotherium* have been quite unsettled. While it is clear that Leidy's "*M.*" *mirabile* belongs in *Eremotherium*, there has been some thought that the North American form and *E. rusconii* (Schaub, 1935), described from Venezuela, represent a single species, as reflected by Kurtén and Anderson's (1980:140) placement of *E. mirabile* in the synonomy of *E. rusconii.* That concept gained some momentum after Gazin (1957) excavated the remains of eight individuals referred to *E. rusconii* at El Hatillo, Panama, the most complete specimens of that species yet found. Commenting on the rather confused state of affairs regarding the two or three described species of *Eremotherium* from tropical America, Gazin (1956:346) suspected that "a single tropical species is represented, as suggested by the Panamanian materials; on the other hand, Paula Couto's allocation of Leidy's *Megatherium mirabile* from Georgia to *Eremotherium* seems valid and the species may well be different. In this case the Georgian species name would be much the oldest pertaining to *Eremotherium.*"

However, Cartelle and De Iuliis (1995) have suggested that *E. mirabile, E. rusconii,* and *E. laurillardi* (Lund, 1842), from Brazil should be synonymized, the lattter name having priority.

Since theirs is the most recent appraisal of the systematics within *Eremotherium,* De Iuliis and Cartelle's (1995) interpretation has been followed here.

Subsequent to the initial discoveries, additional megathere records from South Carolina were published by Leidy (1859:111), Hay (1923:363), and Allen (1926:449), and Roth and Laerm (1980:13) tentatively referred three Charleston Museum specimens (ChM PV2400, a lumbar vertebral fragment, PV2426, an ungual phalanx, and PV2454, a molar fragment) to *E. mirable* (= *E. laurillardi*). All are of Late Pleistocene age.

In late 1974, Floyd Wenningham, a junior high school student near Moncks Corner, Berkeley County, South Carolina, found an astragalus of a large megathere in the fill dirt for an overpass in a section of US. Route 52 then under construction. He took the specimen to his teacher, Mrs. W.H. Wiley, who in turn brought it to The Charleston Museum for identification. The youngster later found that the fill for the overpass had been hauled from a borrow pit on the west side of the new highway 3.9 km southwest of Moncks Corner, and he located the skeletal remains from which the astragalus had been hauled away in fill material. Excavation of the remains was conducted by a Charleston Museum party from 12 May to 19 June 1975.

The skull and mandibles of this animal were missing, but 16 tooth fragments were found. The postcranial elements recovered included the right clavicle, right humerus, right and left radii, right ulna, right second phalanx of the manus, right third ungual of the manus, partial right femur, right and left fibulae, right and left astragali, right tarsal, right fourth metatarsal, right third ungual of the pes, 13 vertebrae, and elements of several ribs. The remains are those of an animal of considerable size, as indicated by the proximodistal length of the right radius (790 mm) and by the third ungual of the pes (Figure 9), which measures 385 mm in anteroposterior length.

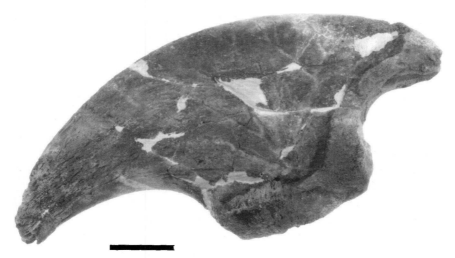

Figure 9. ChM PV4803, third ungual of right pes, *Eremotherium laurillardi*, S.C., Berkeley Co., Ladson Fm. (Middle Pleistocene); left lateral view. Scale bar = 50 mm.

The excavation site borders the valley of a Pleistocene river that apparently was present at the time that the animal was alive. The sediments that contained the bones are clearly fluvial, being a greenish-blue sandy clay overlain by approximately 20 cm of a light-reddish-brown clayey quartz sand that is typical of the Ladson Formation as described by Weems and Lemon (1988). The clay filled a shallow depression approximately 4.4 m in width and 50 cm in depth, evidently the bed of a small stream that once fed into the river. The Late Oligocene Ashley Formation underlies the Ladson Formation at this locality. As was the case with the sloth excavated from the Penholoway Formation (ChM PV4748) and reported in the preceding account, only one or two elements of the left appendages are present, most of the remains being composed of bones from the right limbs, inferring that each animal was lying on its right side after death, thus protecting the right limb elements from scavengers that may have carried off the bones of the left side.

A tooth fragment (ChM PV5842) found by Vance McCollum in the Ten Mile Hill beds in a ditch on Trolley Road (County Road 199), 0.2 mi. (0.3 km) east of South Carolina Route 642 in Dorchester County adds a second Middle Pleistocene record of *Eremotherium* in South Carolina.

* * *

Order CARNIVORA
Family CANIDAE
Genus *Canis* Linnaeus, 1758
CANIS DIRUS LEIDY, 1858
Figures 10, 11

MATERIAL.—ChM PV5756, partial right m1; ChM PV2282, partial right mandibular ramus with root of c1 and p2-4 and m1 preserved (Roth and Laerm, 1980); USNM 437648, skull and mandibles (cast, ChM PV4817).

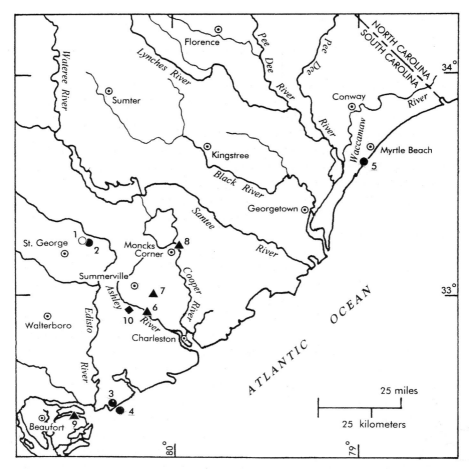

Figure 10. Locality records for *Canis dirus* (○ 1), *Tremarctos floridanus* (● 2-5), *Arctodus pristinus* (π 6-9), and *Ursus americanus* (◆ 10) in South Carolina. **1,** SCSM 91.171.1, 91.171.251-253 (Bentley et al., 1995); USNM 437648; **2,** SCSM 93.105.264-265 (Bentley et al., 1995): Dorchester Co., Giant Cement Company quarry, fissure fill (Late Pleistocene). **3,** ChM GPV2029, GPV2007, PV3463, Charleston Co., Edisto Beach, undetermined offshore unit (Late Pleistocene); **4,** TL1, Charleston Co., Atlantic Ocean, ca. 1.6 km off Edisto Beach, undetermined offshore unit (Late Pleistocene); **5,** AMNH 55539, Horry Co., ca. 4.5 miles (7.2 km) S. Myrtle Beach, Socastee Fm. (Late Pleistocene). **6,** type locality, *A. pristinus,* Charleston Co., Ashley Ferry (Leidy, I854, 1860), Wando Fm. (Late Pleistocene); **7,** ChM PV5146, Berkeley Co., Tall Pines subdivision, Ladson Fm. (Middle Pleistocene); **8,** ChM PV5472, neotype, Berkeley Co., bottom of Tail Race Canal north of U.S. Rt. 52, Wando Fm. (Late Pleistocene); **9,** LH 1, Beaufort Co., bottom of Morgan River, undetermined late Pleistocene unit. **10,** ChM PV2536, Charleston Co., vicinity of Runnymede Plantation, Ashley River; Wando Fm. (Late Pleistocene).

LOCALITY AND HORIZON.—ChM PV5756: S.C., Berkeley Co.; bottom of Cooper River c. 1.0 mile (1.6 km) east of Moreland Landing, c. 15 miles (24 km) north-northeast of Charleston; Ronald Charles, 9 June 1981; Wando Formation, Late Pleistocene. ChM PV2282, S.C., Charleston Co.; Edisto Beach; Edmund R. Cuthbert, Jr., June 1970. USNM 437648: S.C., Dorchester Co.; fissure fill in Giant

Portland Cement Company quarry near Harleyville; Ray Ogilvie, fall 1989; unnamed unit, Late Pleistocene.

AGE.—ChM PV5756, Sangamonian, Rancholabrean; ChM PV2282, USNM 437648, Wisconsinan, Rancholabrean.

DISCUSSION.—Presently known only from the late Pleistocene, the dire wolf has been recorded from Pennsylvania to California and from Alberta, Canada, to Mexico (Nowak, 1979), and a lone record from northern Peru documents its presence in South America as well (Churcher, 1959). *C. dirus* has been reported from South Carolina by Roth and Laerm (1980) on the basis of a partial right mandibular ramus (ChM PV2282) from Edisto Beach and by Bentley et al. (1995), who reported a brain case (SCSM 91.171.1), a left c1 (SCSM 91.171.251), a metapodial shaft (SCSM 91.171.252), and a left jugal (SCSM 91.171.253) from the Giant Portland Cement Company quarry in Dorchester County. Two additional specimens of this taxon from South Carolina are available.

In June 1981 a partial right m1 (ChM PV5756) was found on the bottom of the Cooper River by Robert Charles near the former site of the General Dynamics Company, approximately 1.0 mile (1.6 km) east of Moreland Landing in Berkeley County, some 15 miles (24 km) north-northeast of Charleston. There is no direct evidence of the stratigraphic origin of the specimen, but its probable source can be established from the logs of U.S.Geological Survey auger holes KI 16 (ca. 0.7 mile [1.1 km] west of the bank of the river near the spot at which the *C. dirus* tooth was found) and KI 22 at Moreland Landing (Weems and Lemon, 1989). In the logs of those two auger holes Weems et al. (1985:55, 58) recorded 12 feet of the upper member of the Wando Formation in KI 16 and 13 feet of that unit in KI 22 at Moreland landing, demonstrating that the Wando deposits are continous from auger hole KI 16 eastward to the east bank of the Cooper River adjacent the dire wolf tooth locality. Along that axis the Wando is underlain by the mid-Pliocene Goose Creek Limestone. The upper member of the Wando Formation thus appears to be the only locally available stratigraphic source for this tooth, indicating a minimum age of 60–70,000 years for this specimen. However, the tooth shows appreciable wear and may have been reworked from the middle member of the Wando into the upper member, which would infer an age of approximately 100,000 years. In either event, the specimen documents the presence of *C. dirus* in South Carolina during Sangamonian time.

In the fall of 1989, Ray Ogilvie, an amateur collector from Florence, South Carolina, found the exceptionally well-preserved skull and mandibles (USNM 437648, Figure 11) of a large individual of *C. dirus* in late Pleistocene sediments filling a solution pit in the top of the late Eocene Harleyville Formation in the Giant Cement Company quarry near Harleyville, Dorchester County South Carolina. This locality subsequently furnished the Pleistocene mammal remains that Bentley et al. (1995) have called the Ardis local fauna. Sanders (1974) reported similar cavities in the top of the late Eocene portion of the Santee Limestone that were filled with "Cooper marl" (Harleyville Formation) sediments containing archaeocete remains.

The skull is virtually complete but lacks several teeth. On the right side I1, P1, P2, the posterior half of P3, and M2 are missing. On the left side only I3, C1, P4, and M1 are preserved. The left P3 and M2 seem not to have erupted at all, there being only

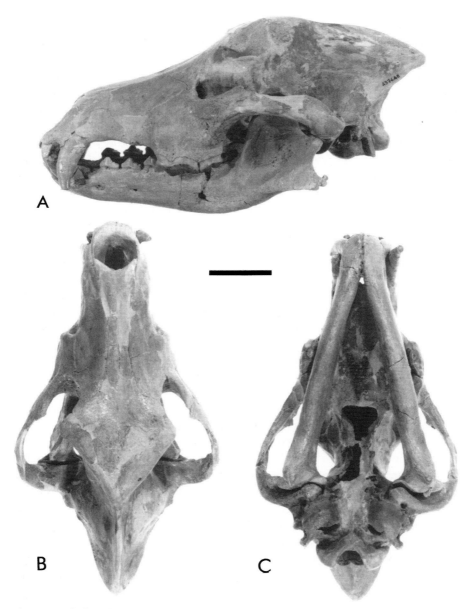

Figure 11. Skull and mandibles of *Canis dirus* (USNM 437648) from late Pleistocene fissure fill in Giant Portland Cement Company quarry near Harleyville, Dorchester Co., S.C. **A**, left lateral view; **B**, dorsal view; **C**, ventral view. Scale bar = 50 mm.

solid bone and no alveolar structures present in the normal locations of those teeth. The mandibles are almost entirely complete, the most notable damage being the absence of the anterior and posterior walls of the right canine alveolus and the lateral wall of the left canine alveolus. With the exception of i1, approximately half of the crown of c1, p1, and the posterior third of m1, the left dentition is intact. The left p1 evidently was lost during life, bone growth having begun to fill the alveolae around two small remnants of the root. The right dentition is missing i1, i2, all but the roots of m2, and m3.

As shown in Table 3, comparison of the measurements of USNM 437648 with those of the largest *C. dirus* skulls measured by Nowak (1979:149) indicates that this specimen ranged near the upper size limits known for dire wolves. Only two skulls examined by Nowak (1979)—a 333-mm specimen from Ingleside, Texas, and a 318-mm specimen from Maricopa, California—are larger than the South Carolina specimen, which measures approximately 317 mm. in greatest length. Although the region around the inion has been restored and thus does not allow an exact measurement of total length, other measurements of USNM 437648 compare favorably with those of the largest specimens studied by Nowak (1979) and suggest that the estimated total length is not significantly different from that of the skull in its original condition. In Table 3 I have used the mean of Nowak's (1979) measurements of the 62 skulls from Rancho La Brea, California, to better place the South Carolina specimen in the context of size variation within a known population. The mean of greatest length of the Rancho La Brea skulls was 294.8 mm, the largest specimen being 316 mm (Nowak, 1979:149). Thus, USNM 437648 probably represents one of the larger individuals within the *C. dirus* population to which it belonged.

* * *

Family URSIDAE
Subfamily TREMARCTINAE
Genus *Tremarctos* Gervais, 1855
TREMARCTOS FLORIDANUS (GIDLEY, 1928)
Figures 10, 12-14, 15a-d

MATERIAL.—AMNH 55539, partial right mandibular ramus with m1-m3 in place; TL1 (cast, ChM PV5757), partial left manibular ramus with m1 and m2 in place; ChM GPV2019, cast of partial right mandibular ramus with m2 in place; ChM GPV2007, right m2 lacking anterior root; ChM PV3463, right M2; DM11, right M2; DM12, posterior portion of left m2; NR2, right M2; NR6, right m1.

LOCALITY AND HORIZON.—AMNH 55539: S.C., Horry Co., ca. 4.5 miles (7.2 km) south of Myrtle Beach, Mrs. N.E. Frost; Socastee Fm.?; Late Pleistocene. TL1: S.C., Charleston Co.; bottom of Atlantic Ocean, ca. 1.0 mile (1.6 km) off Edisto Beach, Terry Lee, October 1991; undetermined offshore unit; Late Pleistocene. S.C., Charleston Co., Edisto Beach: John Bennett (GPV2019), Mr. and Mrs. E. J. Evans (ChM GPV2007), Margaret Pulliam (ChM PV3463), Don and Gracie Marvin (DM11, DM12); undetermined offshore unit; Late Pleistocene.

AGE.— Rancholabrean, Late Pleistocene.

DISCUSSION.—Roth and Laerm (1980:15–16) reported the first record of the North American Spectacled Bear, *Tremarctos floridanus,* from South Carolina, a right mandibular ramus represented by a cast (ChM GPV2019) in The Charleston Museum. Bentley et al. (1995:14) recently reported a right m3 (SCSM 93.105.264)

TABLE 3. Comparison of measurements (in mm) of skull and mandibles of *Canis dirus* (USNM 437648) from Giant Cement Company quarry, Dorchester County, South Carolina, with measurements of *C. dirus* specimens from Rancho La Brea, California, Maricopa, California, and Ingleside, Texas, given by Nowak (1979:149, 153-154). Measurements of USNM 437648 courtesy of Fred Grady (USNM) except where asterisk (*) denotes measurement by the writer to coincide with measurements taken by Nowak (1979). Parentheses ( ) denote estimated measurement. Dashes (—) indicate measurements not available.

	Rancho La Brea, Calif. (mean,n = 62)	USNM 437648	Maricopa, Calif. (1)	Ingleside, Texas (1)
Length, anterior tip of premaxillae to posterior point of inion	294.8	(317.0*)	318.0	333.0
Condylobasal length	—	280.0	—	—
Greatest distance across zygomata	163.3	169.5*	—	179.0
Maximum width of braincase across level of parietotemporal sutures	74.73	80.2*	80.0	79.0
Palate length	—	147.0	—	—
Anterior edge of alveolus of P1 to posterior edge of alveolus of M2	99.99	104.4*	105.0	110.0
Maximum width across postorbital processes of frontals	83.45	88.7*	87.0	—
Anteroposterior diameter at base of crown, C1	15.66	15.0 (lt.)	—	—
Transverse diameter, C1	—	10.0 (lt.)	—	—
Crown length, P4	31.75	32.8 (lt.)	32.5	35.5
" width, P4	—	13.5 *	—	—
Crown length, M1	—	21.1 (lt.)	—	—
" width, M1	—	25.0 (lt.)	—	—
Total length of mandibles	—	214.0	—	—
Anterior edge of alveolus of p1 to posterior edge of alveolus of m3	(mean, n=73) 110.64	112.6 (rt.)*	(mean,n=10) 113.2	118.0
Depth of mandible between p3 & p4 .	31.81	33.9 (rt.)*	34.01	34.0
Anteroposterior diameter at base of crown, c1	—	15.0 (rt.)	—	—
Crown length, p2	—	15.1 (rt.)	—	—
" width, p2	—	6.1	—	—
Crown length, p3	—	16.7 (rt.)	—	—
" width, p3	—	7.2	—	—
Crown length, p4	19.48	20.0 (rt.)	19.87	20.1
" width, p4	—	9.4	—	—
Crown length, m1	34.25	36.1 (rt.)	35.01	36.0
" width, m1	—	12.8	—	—
Crown length, m2	—	13.9 (lt.)	—	—
" width, m2	—	10.0	—	—
Crown length, m3	—	7.2 (lt.)	—	—
" width, m3	—	6.6	—	—

and a right first metatarsal (SCSM 93.105.265) of *T. floridanus* from late Pleistocene sediments in the Giant Cement Company quarry near Harleyville, Dorchester County, South Carolina.

During the summer of 1972 a partial right mandibular ramus with one molar tooth (m2) in place (Figure 12) was found on Edisto Beach, Charleston County,

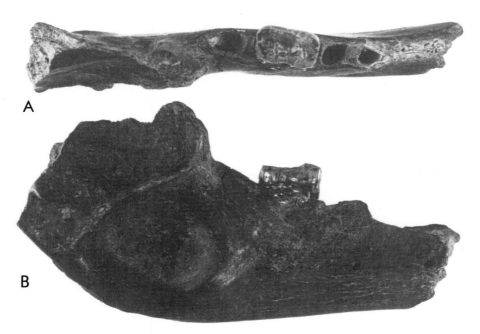

Figure 12. Right mandibular ramus, *Tremarctos floridanus*, with m2 in place. **A**, occlusal view; **B**, labial view. S.C., Charleston Co., Edisto beach; undetermined offshore beds. (Late Pleistocene). Original specimen in private collection; cast, ChM PV2019. × 0.81.

South Carolina (USGS Edisto Island 7.5′ quadrangle), by Mr. John Bennett. The specimen was given to Mr. Charles Harshaw, an amateur fossil collector in Charleston, S.C., who brought it to The Charleston Museum for identification. The specimen was photographed and forwarded to Clayton E. Ray (National Museum of Natural History), who identified it as *Tremarctos floridanus*. Mr. Harshaw elected to keep the specimen, but casts of it were made at the National Museum and are in the vertebrate paleontology collections of that institution (USNM 187131) and The Charleston Museum (ChM GPV2019). Recent attempts to contact that gentleman for re-examination of the mandible were unsuccessful, and the present whereabouts of this specimen are not known. Measurements of the cast (ChM GPV2019) show the preserved portion of this mandible to be approximately 145 mm in anteroposterior length and 47 mm in height at the second molar.

A more nearly complete right mandibular ramus of *Tremarctos floridanus* (AMNH 55539, Figure 13) was found during the summer of 1968 approximately 4.5 mi (7.2 km) south of Myrtle Beach, Horry County, South Carolina (U.S.G.S. Myrtle Beach 15′ quadrangle), by Mrs. N.E. Frost of Houma, Louisiana, who picked it up "on the shore where an old 'swash' flows into the ocean on the south side of Myrtle Beach, just before getting to the airbase, or State Park" (written communication, Mrs. N.E. Frost to R. H. Tedford, AMNH). As in the Edisto Beach mandible (ChM GPV2019), the symphyseal region is missing in the Myrtle Beach specimen (AMNH 55539), but a portion of the coronoid process is preserved in the latter, all three

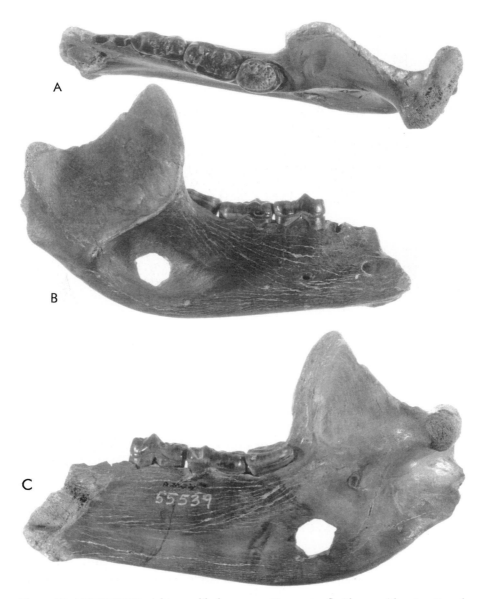

Figure 13. AMNH 55539, right mandibular ramus, *Tremarctos floridanus*, with m1, m2, and m3 in place. **A**, occlusal view; **B**, labial view; **C**, lingual view. S.C., Horry Co.; ca. 4.5 miles (7.2 km) S. Myrtle Beach, Socastee Fm.? (Late Pleistocene). × 0.70.

molar teeth are in place, and the alveolae for the last two premolars are intact. The molars are well preserved and show little wear, indicating that the animal may have been a young adult. The preserved portion of this mandible is 163 mm in antero-posterior length and is 50 mm in height at the second molar.

In October 1991, a left mandibular ramus of *T. floridanus* (TL 1, Figure 14; cast, ChM PV5757) was recovered from the bottom of the Atlantic Ocean at a depth of

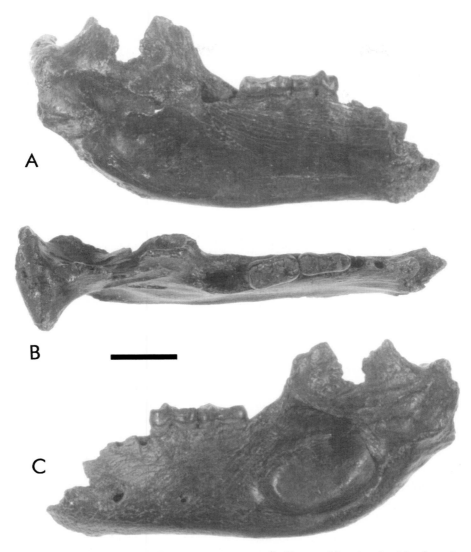

Figure 14. TL 1, left mandibular ramus, *Tremarctos floridanus*, with m1 and m2 in place. **A**, labial view; **B**, occlusal view; **C**, lingual view. S.C., Charleston Co.; bottom of Atlantic Ocean ca.1.0 mile (1.6 km) off Edisto Beach; undetermined offshore unit (Late Pleistocene). Scale bar = 30 mm.

about 55 feet (16.8 m) approximately 1.0 mile (1.6 km) off Edisto Beach by Terry Lee, a local diver. It is missing the angular process, most of the coronoid region, and the anteriormost portion beyond the base of the canine alveolus, but the posterior region of the symphysis is preserved and m1 and m2 are in place. As seen in the list of measurements below, this specimen is considerably larger than the other two rami:

	TL 1	AMNH 55539	ChM GPV2019 (cast)
Anteroposterior length, as preserved	189.0	165.4	145.0
Depth at center of m2	54.1	43.0	47.0
Depth behind m3	64.0	51.0	59.0
Length, m2	23.0	21.5	22.1
Width, m2, anterior	12.7	12.4	12.5
Width, m2, posterior	12.7	12.4	12.6

The measurements of the length of m2 place the two smaller jaws (AMNH 55539 and ChM GPV2019) midway within the observed range (19.0–24.0 mm) for *T. floridanus* as given by Kurtén (1966:13, table 1), the m2 of the largest ramus (TL1) falling into the upper end of the range. The depth of the ramus behind m3 in TL1 (64.0 mm) places this specimen at the maximum end of the range (45.6–64.0 mm) for that measurement in *T. floridanus* rami examined by Kurtén (1966:28, table 5). Based on Kurtén's (1966:86–87) evidence that males of this species are considerably larger than females, the measurements of these three specimens suggest that TL1 is probably the jaw of an adult male, ChM GPV2019 may belong to a young male, and AMNH 55539 may be the jaw of a young female.

Six teeth from Edisto Beach, Charleston County, provide further evidence of *T. floridanus* in South Carolina. A right m2 (ChM GPV2007, Figures 15c-d), found by Mr. and Mrs. E. J. Evans, is intact and well preserved, but only one of the roots is present. A right M2 (ChM PV3463, Figures 15a-b), also missing one of its roots but otherwise well preserved, was found on Edisto Beach by Margaret Pulliam of Charleston about 1980. Another right M2 (DM11), missing two of its roots, and a partial left lower m2 (DM12) were collected on Edisto Beach by Don and Gracie Marvin and are in their private collection. Casts of those teeth are in the U.S. National Museum of Natural History (USNM 475386 and 475385) and The Charleston Museum (ChM PV5715 and PV5716). A right M2 (NR2) and a right m1 (NR6) were found at Edisto Beach by Ned Riddle of Rock Hill, South Carolina, and are in his private collection.

The aforementioned specimens—three mandibular rami and six teeth—constitute the presently known evidence of *Tremarctos floridanus* in South Carolina, and the mandible from Myrtle Beach (AMNH 55539) appears to be the northernmost record of this species on the Atlantic Coastal Plain. An extinct relative of *Tremarctos ornatus*, the modern spectacled bear of Neotropical America, *T. floridanus* is known from Blancan (Pliocene, Plio-Pleistocene) sites in California and Idaho (Kurtén and Anderson, 1980), Irvingtonian (Early-Middle Pleistocene) faunas in California (Downs and White, 1968; Kurtén and Anderson, 1980) and Florida (Webb, 1974a) and from the Upper Pleistocene of Mexico, Texas (Kurtén, 1966), Kentucky (Kurtén and Anderson, 1980), Georgia (Ray, 1967), and Florida (Kurtén, 1966; Martin and Webb, 1974a). A partial skeleton found in a Tennessee cave is of uncertain age (Guilday and Irving, 1967), but *T. floridanus* remains from the Devil's Den fauna of Florida document the presence of this species in Florida as late as 8,000 years ago (Martin and Webb, 1974:138).

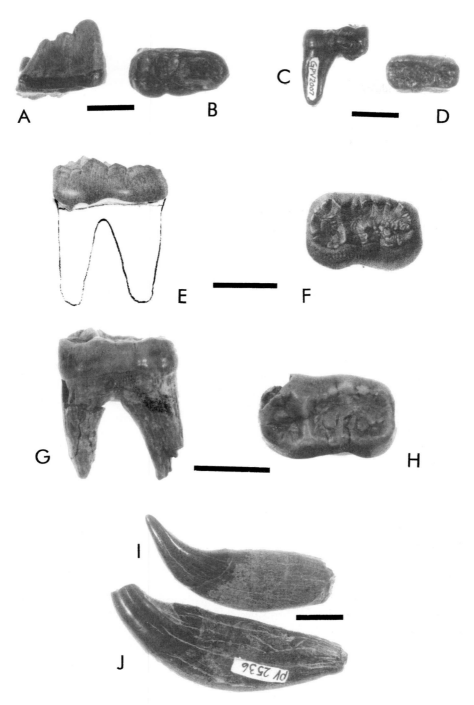

Figure 15. Teeth of *Tremarctos floridanus, Arctodus pristinus*, and *Usus americanus* from South Carolina. **A**, ChM PV3463, *T. floridanus*, right M2, lingual view; **B**, occlusal view; **C**, ChM GPV2007, *T. floridanus*, right m2, labial view; **D**, occlusal view; Charleston Co., undetermined offshore unit (Late Pleistocene). **E**. Holotype left m2, *Arctodus pristinus*, labial view; **F**, occlusal view; (from Leidy, 1860, pl. 23, figs. 3, 4); Charleston Co., Wando Fm. (Late Pleistocene). **G**, ChM PV5146, *A. pristinus*, left m2, labial view; **H**, occlusal view; Berkeley Co.; Ladson Formation (Middle Pleistocene). **I**, ChM PV2567, *Ursus americanus*, lower left C^1, labial view; **J**, ChM PV2536, *Arctodus pristinus*, lower left C^1, labial view; Charleston Co., Wando Fm. (Late Pleistocene). Scale bars = 15 mm.

Having been recorded from middle Pleistocene to post-Wisconianan time, *T. flori-danus* remains alone are of little value in determining the age of Pleistocene beds deposited during that interval. However, the *T. floridanus* specimens from Edisto Beach (ChM PV5757, GPV2007, GPV2019, and PV3463; DM11, DM12) can be reliably assigned to the Late Pleistocene on the strength of evidence provided by other material from Edisto Beach (Roth and Laerm, 1980; see Summary and Conclusions below). Found under similar circumstances, the *Tremarctos* mandible from Myrtle Beach (AMNH 55539) may be of Wisconsinan age. However, it is also possible that that specimen came from the Socastee Formation, in which case it would be of Sangamonian age.

∗ ∗ ∗

Genus *Arctodus* Leidy, 1854
ARCTODUS PRISTINUS LEIDY, 1854
Figures 10, 15e-h, 15j, 16, 17

MATERIAL.—ChM PV5146, left m2; ChM PV5472, partial right mandibular ramus with m1 and m2 in place; ChM PV2536, lower left canine; LH1, distal portion of right radius.

LOCALITY AND HORIZON.—ChM PV5146: S. C., Berkeley Co.; ditch in Tall Pines subdivision, Vance McCollum, August 1981; Ladson Formation, Middle Pleistocene. ChM PV5472: S.C., Berkeley Co., bottom of Tail Race Canal just north of U.S. Route 52, John L. Postek, III, October 1994; Wando Formation, Late Pleistocene. ChM PV2536: S. C., Charleston Co., vicinity of Runnymede Plantation, west side of Ashley River on S.C. Rt. 61, C.C. Pinckney, ca. 1900; Wando Formation, Late Pleistocene LH 1 (cast, ChM PV5718): South Carolina, Beaufort Co., bottom of Morgan River, Lee Hudson, date uncertain; Late Pleistocene, formation undetermined.

AGE.—ChM PV5146: Irvingtonian, Pre-Illinoian. ChM PV2536, ChM PV5472: Rancholabrean, Sangamonian. LH1: Rancholabrean, Sangamonian.

DISCUSSION.—The genus *Arctodus* is represented in North America by *Arctodus simus*, the giant short-faced bear, widely spread across the continent during Irvingtonian and Rancholabrean times, and by *Arctodus pristinus*, the lesser short-faced bear, a smaller form known only from a few Irvingtonian and Rancholabrean localities in the eastern United States (Kurtén and Anderson, 1980:180, fig. 11.13).

Heretofore, *A. pristinus* has been documented in South Carolina only by Leidy's (1854a:90) brief original description of the type, the crown of a left m2 (Figure 15e-f) "discovered by Captain Bowman, U.S.A., in the sands of Ashley river, S.C." (erroneously given as "Ashley River beds, North Carolina" by Kurtén [1967:5]). Kurtén (1967:5) observed that "The age is uncertain but may be middle Pleistocene as for the two dated sites of this species" (i.e. Cumberland Cave, Maryland, and Port Kennedy Cave, Pennsylvania) The holotype of *A. pristinus* is not included in the published account of the fossil vertebrate type specimens at the Academy of Natural Sciences of Philadelphia (Spamer et al., 1995), and recent inquiry has confirmed that the type is not there (E. Daeschler, personal communication, July 1995). Six years after Leidy (1854a:90) described and named the tooth, he stated that it was in "The collection of Captain Bowman" (Leidy, 1860:114) of Charleston, where it apparently remained and was subsequently lost many years ago. In view of that fact, it is puzzling that Kurtén (1967:5) was able to furnish measurements for the specimen, "(length 26 mm, anterior width 18, posterior about 17)," unless he measured the dimensions of Leidy's (1860:pl. 23, figs. 3,4) figures of the tooth. If so, it was an unnecessary step, because measurements of the type do exist. In what may be one of the earliest uses of the metric system in a North American paleontological publication, Leidy (1854a:90) gave the measurements of the tooth as "24 m.m. antero-posteriorly, and 17 m.m. transversely." Hence, those figures are included in Table 4 since they are clearly direct measurements of the now-lost holotype. He later figured the specimen (Leidy, 1860: Pl. 23, figs. 3,4; Figures 15e-f) and noted that it was from the "Post-Pleiocene deposit" at Ashley Ferry on the Ashley River (Leidy, 1860:115–116) "about ten miles north of Charleston" (Leidy,1860:99). Ashley Ferry, also known as Bee's Ferry (Tuomey, 1848:164), appeared on Robert Mills' (1825) map of "Charleston District" (Charleston County) as early as 1820 and was well known as a vertebrate fossil locality, Tuomey (1848:165) remarking on the large numbers of teeth of cartilaginous fishes to be found there. This area is of some significance to the present discussion because it can be established as the type locality of *Arctodus pristinus*, from which point new light can be shed on the probable stratigraphic unit that produced the holotype.

As shown on the Charleston District map in Mills' Atlas (Mills, 1825), Ashley Ferry was located on a now-vanished road that extended from the intersection of present-day Bee's Ferry Road (County Road 57) and Ashley River Road (S.C. Route 61), on the west side of the river, eastward to Dorchester Road (S.C. Rt. 642) on the east side of the river (USGS John's Island 7.5´ quadrangle) (Figure 8). Leidy visited the site with Francis S. Holmes, Curator of The Charleston Museum, in March 1857 and made the following observations:

> Above the Eocene formation there is a stratum of Post-Pleiocene marl, about one foot in thickness, overlaid with about three feet of sand and earth mould. The Post-Pleiocene deposit contains quantities of irregular, water-worn fragments of the Eocene marl-rock from beneath, mingled with sand, blackened pebbles, water-rolled fragments of bones, and other more perfect remains of fishes, reptiles, and mammals, which belong to the Post-pleiocene period, or have been derived from the underlying Eocene formation (Leidy, 1860:99).

The "Eocene" unit beneath the "Post-Pleiocene" stratum is the late Oligocene Ashley Formation, which underlies the entire Charleston area and which was long thought to be of Eocene age (Whitmore and Sanders, 1977) and has at times been reported as Miocene in age (e.g., Kurtén and Anderson, 1980:78). The stratigraphic identity of the one-foot-thick bed of "Post-Pleiocene" sediments uncomformably overlying it at Ashley Ferry can be determined from Holmes' (1858:11; 1860:v) discussion of the molluscan fossils at this locality.

> The fossils from Ashley Ferry present, as a group, the same appearance as those procured inland at some distance from the river, by digging from three

to five feet below the surface. Many specimens from the ferry were considered as recent by Professor Leidy; they appear quite fresh and unchanged in color, and their texture not in the slightest degree altered. . . . it is of common occurrence…among the shells; for example, the olive shell—*Oliva literata*— is found as fresh and highly polished as the recent ones from the sea-beaches along the coast.

The gastropod *Oliva literata* Lamarck, noted above, is now in the synonymy of *Oliva sayana* Ravenel (Campbell et al., 1975). Here, Holmes' mention of "fresh and highly polished" olive shells clearly identifies the Pleistocene deposits at Ashley Ferry. In the Charleston area, fossil specimens of *O. sayana* in that condition, i.e., with the original luster and virtually no wear, are found only in the Wando Formation. This unit is well documented along the west side of the Ashley River above and below the old Ashley Ferry site by two assemblages of mollusks in The Charleston Museum, one from Magnolia Gardens, about 4.9 km northwest of the ferry, and another from a now-filled-in borrow pit 3.3 km southwest of the ferry site. The specimens of *O. sayana* from the borrow pit (ChM PI3200-3298) are exactly as those described by Holmes (1858, 1860), the luster as well preserved as "the recent ones from the sea-beaches along the coast." Thus, it seems clear that the lost holotype tooth of *Arctodus pristinus* came from the Wando Formation, perhaps having been transported by water from an inland source. As noted above, the Wando is of Sangamonian age (c. 87 to 129,000 years [Szabo, 1985]).

A large left lower canine tooth (ChM PV2536,Figure 15j), probably found approximately three miles (4.8 km) north of the type locality, is here referred to *A. pristinus*. Black in color and missing the tip and much of the lingual side of the crown, the tooth is 75 mm in total length as preserved and is 20 mm in transverse diameter at the base of the crown, where a well defined carina arises from a triangular prominence on the lingual surface, as in other canines of *A. pristinus*. The width (20 mm) alone makes this tooth much too large to be a lower canine of either *Tremarctos floridanus* (12.92 mm mean width in Kurtén, 1966:13, table 1) or *Ursus americanus*, the canine of which is dwarfed by comparison with PV2536 (Figures 15i-j). This specimen came to The Charleston Museum in 1957 in the Charles C. Pinckney collection of vertebrate fossils accumulated during his operation of the Magnolia Phosphate Mine and thus seems almost certainly to have come from the Wando Formation.

In August 1981 a left m2 of a tremarctine ursid was donated to The Charleston Museum by Vance McCollum, a Museum volunteer who collected the tooth (ChM PV5146, Figures 15g-h) in a drainage ditch in the Tall Pines subdivision in Berkeley County on the east side of Interstate Highway 26 approximately 27 km northwest of Charleston (USGS Mount Holly 7.5′ quadrangle). The deposits from which the specimen was collected have since been mapped as the Ladson Formation (Middle Pleistocene) by Weems and Lemon (1984b).

The dimensions of this tooth seemed somewhat large for the m2 of *Tremarctos floridanus*, so I sent the specimen to Richard M. Tedford for comparison with AMNH specimens of *Arctodus*. Tedford (personal communication) confirmed it as belonging to *Arctodus*, noting that it is not large enough to be *Arctodus simus* and therefore should be *A. pristinus*. At his suggestion, I compared the tooth with m2 in mandibular rami of *A. pristinus* from the Leisey Shell Pits in Florida from the Florida Museum of Natural History (UF 6400, UF 81694, UF 137957) and the Royal Ontario Museum (ROM 31462) and obtained length/width ratios of those specimens, of ChM PV5146 and PV5472 from South Carolina, and five of the South Carolina specimens of *Tremarctos floridanus* (AMNH 55539,/ChM GPV2007, ChM GPV2019, ChM PV3463, and DM 12). As seen in Table 4, the length/width ratio of ChM PV5146 falls well within those of the other specimens of *A. pristinus* examined during the present study. Kurtén (1967:18) observed that the m2 of *Arctodus* is broader and more robust than that of *Tremarctos*, and that characteristic is reflected in the ratios in Table 4.

Kurtén (1967:27, table 7) reported four partial mandibles for *A. pristinus*, three from Port Kennedy Cave (ANSP 85, 86, 87) and one from Cumberland Cave, Allegany County, Maryland (USNM 8005). He figured a composite restoration of the left mandible of *A. pristinus* based on the three ANSP specimens, but the entire region posterior to the premassetaric fossa is hypothesized by dashed lines, reflecting the incomplete nature of the Port Kennedy specimens (Kurtén, 1967:27, fig. 23). The posterior region of ANSP 87 is fairly complete but is badly crushed (T. Daeschler, personal communication, August 1996). Among the remains of at least eight individuals of *A. pristinus* from the early Pleistocene Leisey Shell Pit Local Fauna in Hillsborough County, Florida, Emslie (1995:503; figs. 2, 3) reported six mandibles and figured four of them. In only one of the figured specimens (UF 81692) was any portion of the coronoid process preserved, and in that specimen only a narrow center section is present (Emslie, 1995:503; fig. 2b). Another Leisey

TABLE 4. Measurements and length-to-width ratios of second molars of *Arctodus pristinus* and *Tremarctos floridanus* examined during the present study. Measurements (in millimeters) are of anterior width except where noted (post. = posteriorwidth; [?]=position of measurement not specified by Leidy [1854]). All are lower molars except where uppers are designated (U). See Materials and Methods for institutional abbreviations. Dashes (—) indicate measurements not available.

Taxon	Specimen	Length	Width	Length into width
A. pristinus	Holotype*	24.0 [?]	17.0 [?]	0.70
"	UF 81694	28.6	18.8	0.66
"	UF 64300	23.2	15.2	0.66
"	ChM PV5146	24.4	15.6	0.64
"	ChM PV5472	27.1	18.4	0.68
"	ROM 31462	23.8	15.5	0.65
T. floridanus	ChM PV5757	23.0	12.7	0.55
"	AMNH 55539	21.5	12.4	0.58
"	ChM GPV2007	21.7	12.1	0.56
"	ChM GPV2019 (cast)	22.1	12.6	0.57
"	DM 12 (incom.)	—	13.2 (post.)	—
"	DM 11 (U)	29.5	15.3	0.52
"	ChM PV3463 (U)	30.0	15.8	0.53

*(From Leidy, 1854a:90)

specimen (ROM 31462), not figured by Emslie (1995), is also missing the coronoid region. Thus, there appear to be no published records of an *A. pristinus* mandible with a well preserved coronoid-condylar region.

In that regard, a partial right mandibular ramus of *A. pristinus* recently donated to The Charleston Museum by Museum volunteer Gary Towles is of particular interest. Although missing m3 and the portion anterior to m1, the remainder of the specimen (ChM PV5472, Figure 16) is complete except for the posterior part of the angular process and a hole in the center of the premasseteric fossa. The measurements of m2 in this specimen (Table 4) fit the observed range in *A. pristinus* as reported by Kurtén (1967:12, table 1) and are comparable to measurements of other specimens of m2 in this species as given in Table 5. ChM PV5472 appears to be the first good evidence of the form of the coronoid and condylar region in *A. pristinus* and confirms Kurtén's (1967:27, fig. 23) reconstruction of that area, especially with regard to the condyle. The position of the condyle was recognized as a diagnostic character of mandibular rami of *A. simus* by Kurtén (1967:28, fig. 23), who stated that "the condyle of *A. simus* is situated so as to place the pivot of the jaw motion exactly in the same plane as the cheek teeth." He further noted that in *T. floridanus* "the condyle is raised high above the plane of the teeth" and that although "the condition in *A. pristinus* cannot be exactly determined . . . there is some suggestion of a slightly raised condyle, as indicated in the tentative restoration" (Kurtén, 1967:28). As seen in Figure 16, the condyle is indeed elevated well above the plane of the cheek teeth in *A. pristinus*, even on its lowest (lingual) side.

One of Kurtén's (1967:27, table 7) standard measurements is the depth of the ramus at the diastema between p2 and p3, but this region is missing in PV5472 so it was necessary to utilize another measurement that allowed comparison of the size of this specimen with that of other mandibular rami of *A. pristinus*. The depth of the ramus at the middle of m2 was selected because that area is preserved in virtually all specimens and thus permits a broad comparison. That measurement was not taken by Kurtén (1967: table 7) or by Emslie (1995) in his study of *A. pristinus* material from the Leisey Shell Pits. Thus, the measurements of mandibular depth at m2 in Table 5 are newly determined figures that provide another means of judging the relative size of mandibular rami and, consequently, the approximate size and probable sex of the animals to which they belong. Sexual dimorphism is common in tremarctine bears, larger individuals usually being regarded as males and smaller ones as females, "although there seems also to be some variation between local or temporal populations" (Kurtén,1967:5, 48). Only a few selected specimens were used in Table 5, the purpose being merely to obtain a general idea of the size of PV5472 in relation to other individuals of *A. pristinus* represented by mandibular rami. Measurements of ROM 31462 from the Leisey Shell Pits, not included in Emslie's (1995) study, are presented here for the first time, along with those of UF 81694, which was figured by Emslie (1995, fig. 2C, D). As seen in Table 5, the depth at m2 in PV5472 (60.0 mm) compares most closely to that of ANSP 87 (60.1 mm), which is the smallest of the *A. pristinus* mandibular rami from Port Kennedy Cave but is still a relatively large individual considering that the largest of the Port Kennedy rami is 70.6 mm in depth at

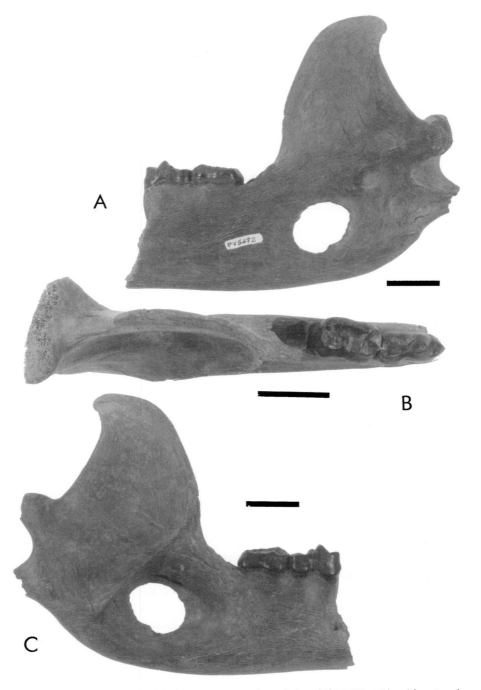

Figure 16. Partial right mandibular ramus, *Arctodus pristinus* (ChM PV5472), with m1 and m2 in place; designated in text as neotype of *A. pristinus*. **A**, lingual view; **B**, occlusal view; **C**, labial view. S.C., Berkeley Co.; bottom of Tail Race Canal; Wando Fm. (Late Pleistocene). Scale bars = 30 mm.

TABLE 5   Measurements (in mm) of mandibular rami and associated second molars of *Arctodus pristinus* from Cumberland Cave, Maryland (USNM 8005), Leisey Shell Pit 1A, Florida (UF 81694, ROM 31642), Port Kennedy Cave, Pennsylvania (ANSP 85, 86, 87), and of the neotype partial mandibular ramus from Berkeley County, South Carolina (ChM PV5472). Asterisks (*) denote measurements from Kurtén (1967:27, table 7). Dashes (—) indicate measurements not available.

	ANSP 85	ANSP 86	ANSP 87	USNM 8005	UF 81694	ROM 31462	ChM PV5472
Depth at diastema	—*	57.5*	52*	c. 63*	—	38.8	—
Depth at middle of m2	70.6	61.0	60.1	67.9	39.3	38.6	60.0
Depth at coronoid process	—*	—*	129*	—*	—	—	145.0
Length, m2	27.0	inc.	25.8	—	28.6	23.8	27.1
Anterior width, m2	13.0	—	12.6	—	18.7	15.3	18.6
Posterior width, m2	13.0	12.0	12.6	—	18.8	15.5	18.1

m2 and the Cumberland Cave specimen (USNM 8005) is 67.9 mm. The anteroposterior length of m2 in PV5472 (27.1 mm) falls about midway in Kurtén's (1967:13, table 1) observed range for the length of this tooth (26.0-28.7). The depth of the ramus at the coronoid process is 145 mm, somewhat greater than Kurtén's (1967:27, table 7) measurement (129 mm) of that point in ANSP 87, with which the South Carolina specimen compares closely in the depth at m2. The South Carolina mandible thus appears to have come from a fairly large individual, perhaps a male. PV5472 is certainly much larger than UF 81694 from Leisey Shell Pit 1A, the latter specimen being only 39.3 in depth at m2 as compared to 60 mm at that point in the South Carolina specimen. However, the length of m2 in the Leisey specimen (28.6 mm) (28.0 in Emslie's [1995:509] table 2) is greater than that of PV5472 (27.1 mm). These two specimens also differ considerably in the position of the condyle and the mandibular foramen. In PV5472 the condyle is elevated well above the level of the teeth, but in UF 81694 it is situated directly on the plane of the cheek teeth, a feature given by Kurtén (1967:28) as a character of *Arctodus simus*. The mandibular foramen is on a level with the crowns of the teeth in PV5472, but in UF 81694 it is on a level with the roots of the teeth, consistent with the position of the condyle. The length of the m2 in UF 81694 is exceptionally great for this relatively small individual, exceeding the length of m2 in the largest of the Port Kennedy mandibular rami of *A. pristinus* ((27.0 mm, ANSP 85), just barely fitting into the uppermost level of Kurtén's (1967:12, table 1) observed range of the length of m2 in *A. pristinus* (26.0-28.7) but fitting well within his observed range for *A. simus* (26.4-33.6) (Kurtén, 1967:13, table 1). The m1 of ChM PV5472 is 28.2 mm in length, placing it in the upper size range of *A. pristinus* (27.3-28.8) (Kurtén, 1967:13, table 1). Both Emslie's (1995:509, table 2) measurement of the length of m1 in UF 81694 (29.5 mm) and my own measurement of that tooth (30.3 mm) place it closer to the range given by Kurtén (1967:12, table 1) for *A. simus* (29.9-35.3) than to that given for *A. pristinus* (27.3-28.8). Another specimen from Leisey Shell Pit 1A referred to *A. pristinus* by

Emslie (1995:508, table 1) seems somewhat large for that taxon. The length of UF 67089, an M2, given as 39.0 mm in length, is well beyond Kurtén's (1967:12, table 1) size range for M2 in *A. pristinus* (35.5-37.3) but fits quite comfortably within his range for M2 in *A. simus* (33.3-41.4).

In view of its exceptionally good preservation, not only of the bone but also of the m1 and m2, but particularly because of its stratigraphic origin (almost certainly the same formation as that of the long-lost holotype), the partial right mandibular ramus maintained as ChM PV5472 is here designated as the neotype of *Arctodus pristinus* Leidy, 1854.

The first postcranial element of *A. pristinus* known from South Carolina, a partial right radius (Figure 17), is in the private collection of Lee Hudson of Florence, South Carolina, and a cast of it is in The Charleston Museum (ChM PV5718). The specimen compares quite favorably with radii of the specimen of *Arctodus pristinus* from Cumberland Cave (USNM 8005) (personal communication, Frederick Grady, National Museum of Natural History, April 1996). Hudson found the specimen on the bot-

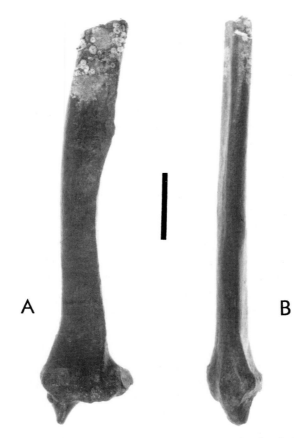

Figure 17. Right radius, *Arctodus pristinus* (LH 1), missing proximal end. **A**, anterior view; **B**, lateral view; **C**, medial view. S.C., Beaufort Co., bottom of Morgan River; Wando Fm.? (Late Pleistocene). Scale bar = 50 mm.

tom of the Morgan River in Beaufort County, South Carolina (Figure 10), in sediments of Late Pleistocene age, as indicated by elements of other Rancholabrean taxa that he recovered from the same matrix (e.g., *Casteroides, Mammut,* and *Mammuthus*). The exact age of this underwater deposit has not been determined, but the locality is approximately the same distance inland and is of comparable elevation as the Wando Formation elsewhere, indicating that the deposit is probably the Wando Formation or is a Wando-equivalent unit and thus is of Sangamonian age. Missing approximately 13.5% of its proximal end, the specimen measures 296 mm as preserved. Comparison with radii of the black bear, *Ursus americanus,* in the Recent mammal collection of The Charleston Museum suggests that this bone was approximately 342 mm in length in its complete state. Measurements of the specimen (in mm) are as follows, in certain instances quoting Kurtén's (1967:32, table 11) wording to facilitate comparison with his measurements of *Arctodus* radii:

Length of specimen as preserved	296
Estimated length in complete state	342
Anteroposterior diameter, approximate middle of diaphysis	28.8
Transverse diameter, approximate middle of diaphysis ("middle of shaft, short diameter")	20.8
Anteroposterior diameter, distal end ("distal end, long diameter")	67.1
Transverse diameter, distal end ("distal end, short diameter")	39.1

The new material reported herein provides a fairly accurate record of the temporal distribution of *A. pristinus* in South Carolina as presently understood. As noted above, the deposits from which the Berkeley County m2 (ChM PV5146) was collected have been mapped as the Ladson Formation (Middle Pleistocene) by Weems and Lemon (1984b). Consisting of marine and marginal marine sediments, this formation has been correlated with the Canepatch Formation, a 450,000-year-old unit at Myrtle Beach, South Carolina (McCartan et al., 1990:A16). Thus, on the basis of the lost holotype (Leidy, 1854a, 1860) and the newer specimens (ChM PV5146 and PV5472), it appears that *A. pristinus* occurred in South Carolina at least from late Irvingtonian to middle Rancholabrean time, c. 450,000 to 87,000 years ago. That span fits the known temporal distribution of *A. pristinus* as defined by Kurtén and Anderson (1980:180), who stated that "Its known history . . . ranges from the early Kansan to the Wisconsinan, when it seems to have been a relict confined to Florida." Its presence further north in present-day South Carolina suggests that *A. pristinus* may have ranged throughout the southeastern United States. Establishment of the stratigraphic origin of the holotype from the Ashley River beds near Charleston now indicates that *A. pristinus* occurred in South Carolina as late as the Sangamonian interglacial period, and it may have lingered on into the Wisconsinan. To date, however, *Tremarctos floridanus* is the only tremarctine ursid definitely known from the Wisconsinan in South Carolina.

Subfamily URSINAE
Genus *Ursus* Linnaeus, 1758
URSUS AMERICANUS PALLAS, 1780
Figures 10, 15i

MATERIAL.— A virtually complete lower left canine (ChM PV2567).

LOCALITY AND HORIZON.— South Carolina; locality uncertain but probably Charleston County, vicinity of Runnymede Plantation, S.C. Route 61, ca. 11 miles northeast of Charleston. Wando Formation, Late Pleistocene.

AGE.— Rancholabrean, Sangamonian.

DISCUSSION.— Hay (1923a:363) included *Ursus americanus* in his list of Pleistocene mammals from South Carolina but did not cite a published source and did not discuss the specimen on which he based the record, indicating only that it was from "somewhere around Charleston." As seen in Table 3, there appear to be no subsequent published records of this bear from the Pleistocene of South Carolina, seemingly a rather strange circumstance in view of the fact that black bears are still a part of the mammal fauna of South Carolina, though they are now greatly reduced in numbers and are restricted to the larger Coastal Plain river swamps and the mountainous regions of the state. There are, in fact, more fossil remains of extinct tremarctine bears from South Carolina than there are of *U. americanus*, so it seems appropriate to call attention to ChM PV2567 (Figure 15i). This tooth, a well preserved left lower canine, is missing only the distal tip of the root. A part of the Pinckney collection, it was one of the specimens that Hay examined during his visit to The Charleston Museum in 1915. In the files of the Department of Paleobiology at the National Museum of Natural History there is an unpublished manuscript in which Hay intended to report some of the specimens that he had seen in Charleston, and in a brief account under the heading "*Ursus americanus*" he noted the presence of a left lower canine in the Pinckney collection. Thus, there seems to be little doubt that ChM PV2567 is the specimen upon which Hay's (1923a:363) record of this taxon is based.

The stratigraphic origin of this specimen is not known, but, as observed above in the discussion of the *A. pristinus* canine (ChM PV2536), its black coloration indicates that it probably came from the lag deposit at the base of the Wando Formation. As reported above, there are seven records of *Tremarctos floridanus* and four records of *Arctodus pristinus* known from the Pleistocene of South Carolina, a total of 11 specimens of tremarctine bears compared to the single specimen of *Ursus americanus* from the state. That disparity suggests that *U. americanus* did not compete well with the larger tremarctines and perhaps did not invade the regions south of Canada until relatively late in the Pleistocene. The pattern of a greater preponderance of tremarctines over ursines is seen elsewhere in the fossil record of North America (Richard E. Tedford, personal communication, 4 February 1997), and Webb (1974a, table 2.1, Fig. 2.2) indicates that the first evidence of that species in Florida is from approximately late Sangamonian time (Arredondo IA and IIA). Even in the Devil's Den fauna of Florida, estimated to be of latest Wisconsinan or early Holocene age, "Specimens of *Ursus* occur less commonly than those of *Tremarctos*" (Webb, 1974a:14, 129).

Family FELIDAE
Subfamily MACHAIRODONTINAE
Genus *Smilodon* Lund, 1842
SMILODON FATALIS (LEIDY, 1868)
Figures 18, 19, 20

MATERIAL.—A left P4 (DM 2) and a fragment of a left maxilla with P4 and the alveolus for P3 (RBH 1; cast, ChM PV3002).

LOCALITY AND HORIZON.— DM 2, : S.C., Charleston Co., Edisto Beach near Jeremy Inlet; Mary Waters, 5 September 1991. RBH 1 (casts, USNM 482285, ChM PV3002): S.C., Charleston Co, Edisto Beach; Betty Harkless, February 1983. Undetermined offshore unit. Late Pleistocene.

AGE.—Rancholabrean, Wisconsinan.

DISCUSSION.—Sabertooth remains are known from numerous Rancholabrean sites in Florida (Webb, 1974a) and elsewhere, but the only published record of *S. fatalis* from South Carolina is based on a left occipital condyle from the Ardis local fauna in Berkeley County tentatively referred to that taxon by Bentley et al. (1995:10). Thus, the specimens noted above are the first definite records of *Smilodon* in South Carolina, and, according to Kurtén and Anderson (1980:187, fig. 11.17), this genus has not been previously documented on the Atlantic Coastal Plain north of Florida.

DM 2 is a well preserved left P4 (Figure 19). Well-worn occlusal facets are present on both crests of the crown but are not so advanced as to suggest that the tooth is that of an old individual. This specimen is in the private collection of the late Don Marvin of Edisto Beach.

RBH 1 is a fragment of left maxilla with the alveolus for P3 and with P4 in place but missing virtually all of the occlusal region of P4 (Figure 18). Shortly after its discovery in 1983, the collectors, Ray and Bettie Harkless of Transfer, Pennsylvania, brought the specimen to The Charleston Museum for identification. It was identified as *Smilodon* by the writer and verified by Clayton E. Ray of the National Museum of Natural History, where casts were made for the National Museum (USNM 482285) and the Charleston Museum (ChM PV3002) collections. At the time of its discovery, this specimen was the first record of *Smilodon* from South Carolina, and at present it remains in the hands of the collectors.

Measurements (in mm) of the original specimens are as follows:

	DM 2	RBH 1
Anteroposterior length of P4 at base of crown	38.8	36.8
Greatest height of crown, P4	22.3+	15.3+
Greatest vertical diameter of P4 as preserved (level of tips of roots to tip of crown)	48.0	36.7+
Anteroposterior length, alveolus for P3	—	12.9
Anteroposterior length of maxillary fragment, as preserved	—	57.3

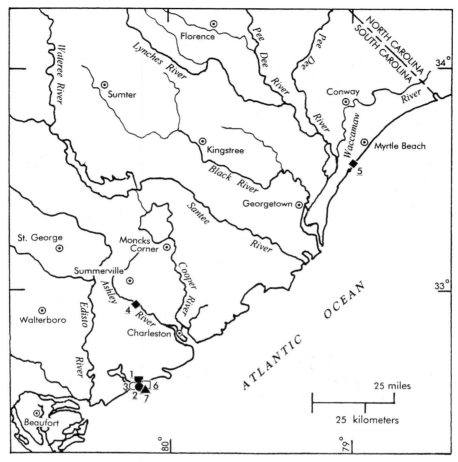

Figure 18. Localities for *Smilodon fatalis* (■ 1), *Panthera leo atrox* (● 2), *Panthera onca augusta* (○ 3), *Miracinonyx inexpectatus* (◆ 4-5), *Puma concolor* (□ 6), and *Lynx rufus* (▲ 7) from South Carolina reported in the present paper. **1**, DM 2; RBH 1 (cast, ChM PV3002): Charleston Co., Edisto Beach, undetermined offshore unit (Late Pleistocene). **2**, NR 1 (cast, ChM PV5929), Charleston Co., Edisto Beach, undetermined offshore unit (Late Pleistocene). **3**, ChM PV2284 (Ray 1967), ChM PV5755; casts, USNM 489152, ChM PV4891: Charleston Co., Edisto Beach, undetermined offshore unit (Late Pleistocene). **4**, MCZ 16512, "phosphate beds of Ashley River" (Allen, 1926:448), approximate location, Charleston Co., Penholoway Fm. (Early Pleistocene); **5**, USNM 299910, Horry Co., Surfside Beach, Waccamaw Fm. (Early Pleistocene) **6**, DM3, Charleston Co., Edisto Beach, undetermined offshore unit (Late Pleistocene). **7**, ChM PV3462, Charleston Co., Edisto Beach, undetermined offshore unit (Late Pleistocene).

The smaller size of P4 in RBH 1 compared to DM 2 (i.e. a difference of 3.3 mm in width) infers that this specimen might be referable to *S. gracilis* Cope, 1880, a poorly-known early Irvingtonian form that was somewhat smaller than *S. fatalis* (Kurtén and Anderson, 1980:186). However, the almost certain absence of Irvingtonian-age beds off Edisto Beach indicates that RBH 1 is more likely the tooth of a small individual of *S. fatalis*, perhaps a young animal, so it is therefore referred to that species.

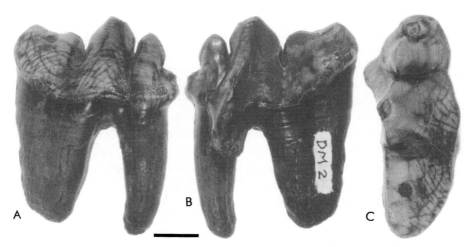

Figure 19. Left P4, *Smilodon fatalis* (DM2). **A**, lingual view, **B**, labial view, scale bar = 10 mm; **C**, occlusal view, scale bar = 5 mm. S.C., Charleston Co., Edisto Beach, undetermined offshore unit (Late Pleistocene).

The genus *Smilodon* has received considerable nomenclatural attention within the past 25 years. Webb (1974b) concluded that *Smilodon californicus* Bovard, 1907, was conspecific with *S. floridanus* (Leidy, 1889) and that *S. floridanus* was distinct from *S. fatalis* (Leidy, 1868). Kurtén and Anderson (1980) placed all three taxa within the synonymy of *S. fatalis* (Leidy, 1868), which has priority. Subsequently, Berta (1985) synonymized *S. fatalis* with the Late Pleistocene South American species *S. populator* Lund, 1842, but Kurtén and Werdelin (1990) have presented convincing evidence that *S. fatalis* is a valid species, and their interpretation has been followed in the present paper.

* * *

Subfamily FELINAE
Genus *Panthera* Frisch, 1775
***Panthera leo atrox* (Leidy, 1853)**
Figures 18, 21

MATERIAL.—Fragment of right maxilla with P4 and posterior half of P3 (NR 1; casts, USNM 489153, ChM PV5929).

LOCALITY AND HORIZON.—S.C., Charleston Co., Edisto Beach; Mrs. Ned Riddle; date not recorded. Undetermined offshore unit; Late Pleistocene.

AGE.—Rancholabrean, Late Wisconsinan.

DISCUSSION.—Larger than the modern African lion, the American lion has been recorded from Sangamonian to late Wisconsinan deposits at numerous localities from Alaska to Wyoming, Arizona, Kansas, Missouri and Texas, but in the southeastern United States it has been reported only from Florida (Kurtén, 1965; Kurtén and Anderson, 1980). The present specimen, a fragment of the right maxilla bearing a well-preserved P4 and the posterior half of P3 (Figure 19), was found several years ago at Edisto Beach, Charleston County, by Mrs. Ned Riddle and is now in the private collection of her son, Ned Riddle, of Rock Hill, South Carolina. Not long after he received the specimen, Mr. Riddle took it to the Schiele Museum of Natural History in Gastonia, North Carolina, where casts were made of it. A cast was placed in the National Museum of Natural History (USNM 489153), where the jaw fragment was determined as *Panthera atrox*. In July of 1997 Mr. Riddle kindly made the original specimen available to me for study. It is of considerable significance because it appears to be the first evidence of *P. leo atrox* in South Carolina. The specimen preserves the portion of the maxilla extending from a point just anterior to the anterior root of P3 posteriorly to the exposed posterior root of P4. Approximately 90 per cent of the infraorbital foramen is preserved, along with a small portion of the jugal forming the lower anterior margin of the orbit. The suture of the jugal and the maxilla is barely discernable, suggesting that the animal was an individual well into adulthood.

Measurements (in mm) of this specimen as preserved are as follows:

Anteroposterior length (as preserved)	63.3
Greatest vertical diameter	74.7
Anteroposterior length of P4	34.1
Anterior width of P4	18.9

Merriam and Stock (1932) give the ranges of size of P4 in the American lion as 35-45 mm in anteroposterior length and 18.3-22.9 mm in anterior width. Kurtén (1973) gives the size ranges of P4 in the jaguar (*Panthera onca augusta*) as 26.8-33.5 mm in anteroposterior length and 14.6-16.7 mm in anterior width. Among unpublished measurements of 32 specimens of P4 in *Panthera onca*, the largest (USNM 23486 from Ladds, Georgia [not examined by Kurtén]) was 33.9 mm in anteroposterior length and 16.8 mm in anterior width (Kevin Seymour, personal communication, August, 1997). The measurements of the length of P4 in the Edisto Beach specimen (NR1) place it near the boundary of *P. onca* and *P. l. atrox*, but its anterior width (18.9 mm) would seem to exclude it from consideration as *P. onca* and to place it clearly within the ranges for *P. l. atrox* according to Merriam and Stock (1932). The relatively small size of the tooth and the poorly defined suture of the jugal and the maxilla indicate that the animal may have been an old female of moderate size.

* * *

PANTHERA ONCA AUGUSTA (LEIDY, 1872)
Figures 18, 22

MATERIAL.—ChM PV2284, right P4. USNM 489152 (cast), ChM PV4891 (cast); anterior end of right maxilla with base of canine and alveolae for I3, P3, and P4. ChM PV5755, right third metacarpal.

LOCALITY AND HORIZON.—S.C., Charleston Co.; Edisto Beach; undetermined offshore unit; Late Pleistocene: ChM PV2284, Gerald Case, 1966; casts, USNM 489152, ChM PV4891, of specimen collected by Rob Thomas, 1985; ChM PV5755, Don and Gracie Marvin, ca. 1992.

AGE.—Rancholabrean, Late Wisconsinan.

DISCUSSION.—This large Nearctic jaguar was first reported from South Carolina by Ray (1967:136–137), who called attention to an isolated right P4 (ChM PV2284, Figure 22a-b) from Edisto Beach in the Charleston Museum collection. Kurtén (1973) and Roth and Laerm (1980) have also made note of this specimen, collected by Gerald Case in 1966. It is 30.8 mm in anteroposterior length, and the transverse diameter of the base of the crown is 16.0 mm at its greatest width anteriorly.

More recently, a fragment of a right maxilla, found on Edisto Beach sometime in 1985 by Rob Thomas, an amateur collector, was identified as *Panthera onca augusta* at the National Museum of Natural History. A cast of the fragment as preserved (ChM PV4891, Figure 22c-d) measures 74.2 mm in anteroposterior length, 66.2 mm in height, and 30 mm in greatest transverse diameter. The distal portion of the canine tooth is broken off 5 to 10 mm below the margins of the alveolus, and at that point the root is 24 mm in anteroposterior length and 19 mm in transverse diameter, corresponding almost exactly to the measurements (24.2 and 19.8 left, 24.0 and 18.5 right) given by Ray (1967:134–137) in his report of an associated skull and

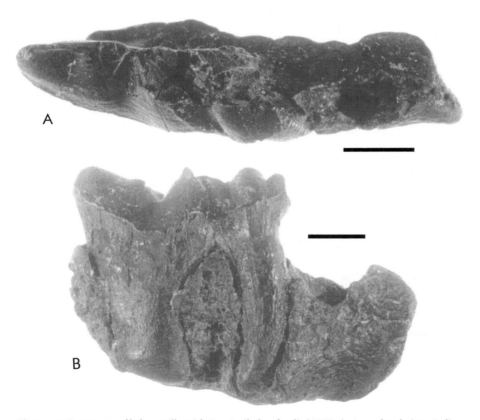

Figure 20. Fragment of left maxilla with P4, *Smilodon fatalis* (RBH 1). **A**, occlusal view; **B**, lingual view. S.C., Charleston Co., Edisto Beach, undetermined offshore unit (Late Pleistocene). Scale bars = 10 mm.

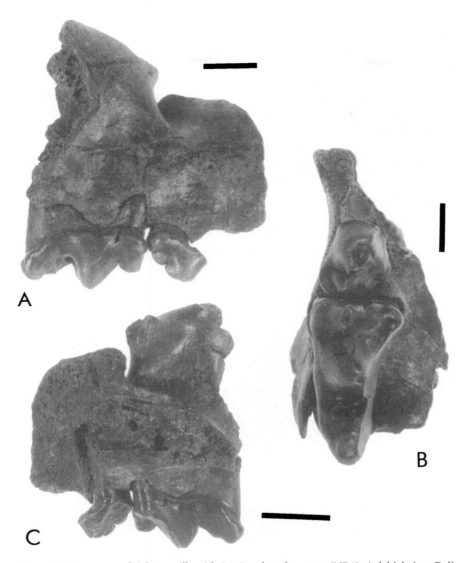

Figure 21. Fragment of right maxilla with P4, *Panthera leo atrox*, (NR1). **A**, labial view; **B**, lingual view; **C**, occlusal view. S.C., Charleston Co., Edisto Beach, undetermined offshore unit (Late Pleistocene). Scale bars = 15 mm.

mandibles of *P. onca augusta* (USNM 23486) from late Pleistocene fissure fillings at Ladds, Bartow County, Georgia.

A right third metacarpal (ChM PV5755, Figure 22e) from Edisto Beach is the first known postcranial element of *P. onca augusta* from South Carolina. Identified by Frederick Grady (National Museum of Natural History), the specimen measures 86.3 mm in length, close to the length of 90 mm cited by Kurtén (1965:259, table 13) for each of two third metacarpals from the late Pleistocene Reddick site in Florida.

To date, the isolated P4 (ChM PV2284), the maxillary fragment (USNM 489152, ChM PV4891), and the metacarpal (ChM PV5755) appear to be the only records of

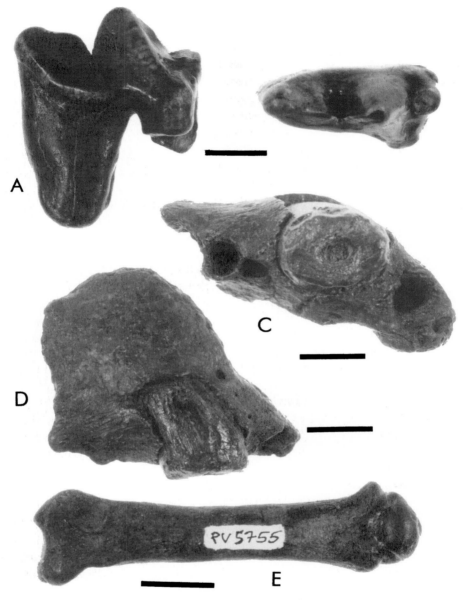

Figure 22. *Panthera onca augusta* remains from Edisto Beach, Charleston Co., S.C.; undetermined offshore unit (Late Pleistocene). **A**, ChM PV2284, right P4 (Ray, 1967), labial view; **B**, occlusal view; scale bar = 10 mm. **C,** ChM PV4891, cast of anterior end of right maxilla, occlusal view; scale bar = 15 mm; **D**, labial view; scale bar = 20 mm. **E**, ChM PV5755, right 3rd metacarpal, dorsal view; scale bar = 15 mm.

jaguars in the Pleistocene of South Carolina. Elsewhere, Pleistocene records of *P. onca augusta* in North America are widely distributed from Florida to Maryland, westward to New Mexico and Washington (Kurtén and Anderson, 1980:192), and as far north as Lost Chicken Creek in Alaska (USNM 23619, 262536). Approximately 15–20% larger than the modern jaguar (*P. o. onca*), *P. onca augusta* undewent a reduction in both its size and geographic range during the middle and late Pleistocene (Kurtén and Anderson, 1980; Seymour, 1993).

* * *

Genus *Miracinonyx* Adams, 1979
MIRACINONYX INEXPECTATUS (COPE, 1895)
Figures 18, 23

MATERIAL.—MCZ 16512, left mandibular ramus with roots of p3, p4, and m1 but missing anterior and posterior ends of ramus; USNM 299910, right mandibular ramus with roots of p3, p4, and m1 but missing anterior and posterior ends of ramus.

LOCALITY AND HORIZON.—MCZ 16512: S.C., Charleston Co.; phosphate mine on Ashley River; Pringle Frost, c. 1900; Penholoway Formation, Early Pleistocene. USNM 299910: S.C., Horry Co., Surfside Beach; R.E. Smith; Waccamaw Formation, Early Pleistocene.

AGE.—MCZ 16512, Middle Irvingtonian, Pre-Illinoian; USNM 299910, Early Irvingtonian, Pre-Illinoian.

DISCUSSION.—Allen (1926:448) reported a partial left mandibular ramus (MCZ 16512) from "The phosphate beds of Ashley River, South Carolina" and referred it to "*Felis* sp. (?*cougar* Kerr)," noting that this specimen, "though lacking the crowns of the teeth, is practically identical in size with that of the living Puma of the eastern United States" (*Felis cougar* Kerr, 1792, as used by Allen; *Puma concolor* of current usage). Nevertheless, the fragmentary nature of the specimen (Figure 23a, c) and an uncharacteristic disparity in the size of the roots of the third premolar dissuaded Allen from a more confident assignment to *Puma concolor*. As seen in Figure 23a, the anterior root of p3 is considerably smaller than the posterior one, Allen (1926:448) giving the transverse diameter of the anterior root as 3.8 mm and that of the posterior root as 5.4 mm. Allen (1926:448) observed that "Although no other modern specimen [of *Puma concolor*] has been seen that equals this disparity between the two roots, it seems more likely that it is a matter of individual variation than of specific difference." There is, however, such a consistent similarity in the sizes of the two roots of the p3 in *Puma concolor* that there seems to be virtually no chance that MCZ 16512 is referable to that species, as illustrated by DM 3, a right mandibular ramus of *Puma concolor* reported below from Edisto Beach, Charleston County, South Carolina. In that specimen (Figure 24b) the anterior root of p3 is actually slightly longer (5.7 mm) anteroposteriorly than the posterior root (5.4 mm). In MCZ 16512 the alveolus of the anterior one measures 4.7 mm in anteroposterior length and the posterior one is 6.5 mm in length. Transversely, the anterior root in DM 3 (*P. concolor*) measures 4.3 mm and the posterior one 4.4 mm, much closer proportionately than Allen's (1926:448) measurements of 3.8 mm (anterior) and 5.4 mm (posterior) in MCZ 16512, which were verified during examination of the specimen in connection with the present study. The size, shape, and orientation of the anterior root in that specimen is at such variance with that of DM 3 that *P. concolor* can be disregarded as a viable candidate for the assignment of MCZ 16512.

There are other felids that must be considered. In the dentary of *Panthera onca augusta* the anterior root of p3 is much smaller than the posterior one, as in MCZ

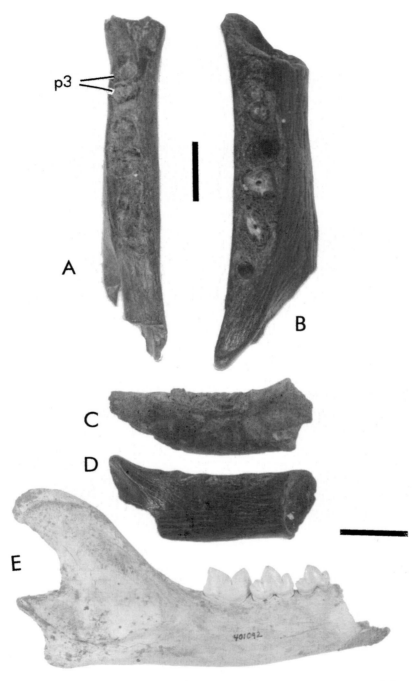

Figure 23. Mandibular rami of *Miracinonyx inexpectatus.* **A**, MCZ 16512, left mandibular ramus, occlusal view; "phosphate beds of Ashley River" (Allen, 1926), S.C., Charleston Co., Penholoway Fm. (Early Pleistocene); **B**, USNM 29910, right mandibular ramus, occlusal view; S.C., Horry Co., Waccamaw Fm. (Early Pleistocene); scale bar = 15 mm. **C**, MCZ 16512, labial view; **D**, USNM 29910, labial view; **E**, USNM 401092, right mandibular ramus, labial view; West Virginia, Franklin Co., Hamilton Cave (Middle Pleistocene); scale bar = 30 mm.

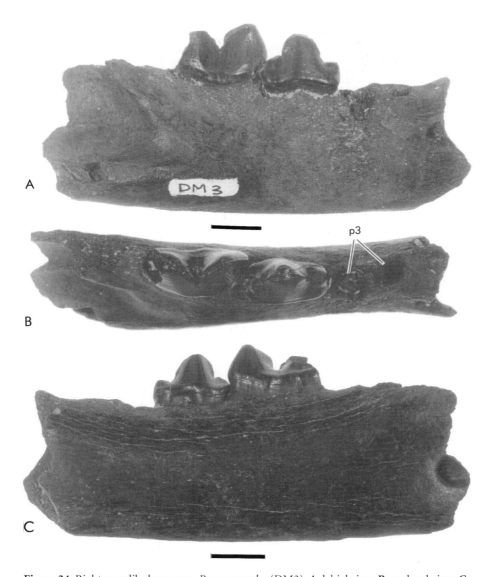

Figure 24. Right mandibular ramus, *Puma concolor* (DM3). **A**, labial view; **B**, occlusal view; **C**, lingual view. S. C., Charleston Co., Edisto Beach, undetermined offshore unit (Late Pleistocene). Scale bars = 10 mm.

16512, but the latter specimen is too small to be a jaguar jaw; i.e., the depth of the ramus behind m1 is 27.8 mm compared to a range of 37.8 to 44.4 mm in four adult specimens of *P. o. augusta* from the Pleistocene of Florida (Kurtén, 1965:255, table 5). Some of the measurements of MCZ 16512 are comparable to those of a juvenile mandibular ramus of *P. o. augusta* (UF 8455) from Florida given by Kurtén (1965:254, table 4), but the roots preserved in the alveolae of the South Carolina specimen are clearly those of the adult dentition.

Comparison of MCZ 16512 with the partial right mandible preserved with USNM 401092, a nearly complete skeleton of the cheetah-like cat *Miracinonyx inexpectatus* from Hamilton Cave, West Virginia, leaves virtually no doubt that the South Carolina specimen is referable to *Miracinonyx*. In the right mandible of the Hamilton Cave individual the anterior root of p3 is decidedly smaller than the posterior one, as in MCZ 16512 . The length of the alveolar row (anterior margin of p3 to posterior margin of m1) in the mandible of USNM 401092 is 51.8 mm (Van Valkenburgh et al., 1990:440) and in MCZ 16512 it is 46.9 mm, the latter measurement being well within the range of measurements cited by Van Valkenburgh et al. (1990:440) for USNM 401092 (*Miracinonyx inexpectatus*), the Rancholabrean form *Miracinonyx trumani* (44.5 mm), the extinct Old World cheetah *Acinonyx pardinensis* (52.8 mm), and the living *Acinonyx jubatus* (42.7 mm).

A felid partial right mandible (USNM 299910, Figure 23e) from Surfside Beach, Horry County, South Carolina, catalogued in the USNM collection as "*Felis* sp.," is here referred to *Miracinonyx* on the same grounds as MCZ 16512. Ranging between USNM 401092 and MCZ 16512 in size, missing its anterior and posterior ends, and considerably more worn than the latter specimen, it nontheless preserves the roots of p3, p4, and m1. As in the other two specimens, the anterior root of p3 is smaller than the posterior one, and, as seen below, the other measurements of this specimen are proportionately comparable to those of USNM 401092 and MCZ 16512.

	MCZ 16512	USNM 299910	USNM 401092
Greatest anteroposterior length of ramus as preserved	86.0	87.7	152.5
Alveolar length, p3-m1	46.9	47.6	51.8
Transverse diameter, anterior root of p3	3.8	4.8	5.2
Transverse diameter, posterior root of p3	5.4	5.4	6.5
Depth of ramus at center of p3, lingual side	27.4	28.5	30.6
Depth of ramus behind m1	27.8	(broken)	31.6
Thickness of ramus below m1	12.0	13.5	14.0

These measurements place MCZ 16512 and USNM 299910 quite out of the size range of the smaller felid species such as *Herpailurus yagourundi* (Jaguarundi), *Felis pardalis* (Ocelot) and *Lynx rufus* (Bobcat), as well as that of the larger cats *Panthera leo* and *Panthera onca*. The measurements are compatible with those of *Puma concolor*, a possible relative of *Miracinonyx* (Van Valkenburgh et al.,1990), but the disparity in the size of the roots of p3 mitigate against referral of the two South Carolina specimens to *P. concolor*.

Morphometrically, both of the South Carolina specimens compare favorably with the mandibular ramus of *Miracinonyx inexpectatus* from Hamilton Cave (USNM 401092) and are here referred to that form.

The earliest record of the genus *Miracinonyx* is the Blancan form *M. studeri* from Cita Canyon, Texas (2.6 Ma [Repenning, 1987]) and from California. *M. inexpectatus* is known from the late Pliocene (Hulbert, 2001), from the early Pleistocene Inglis IA and Leisey Shell Pit 3 sites in Florida, both of which are approximately 1.7 Ma in age (Hulbert, 2001), and from Irvingtonian sites in California, Nebraska, Texas, Arkansas (Conard Fissure, c. 0.7 Ma [Repenning, 1987]), Maryland (Cumberland Cave), Pennsylvania (Port Kennedy Cave), and West Virginia (Hamilton Cave), the latest known occurrence of *M. inexpectatus* having been dated at about 0.6 Ma. The Rancholabrean form, *M. trumani*, has been recorded from Crypt Cave, Nevada (ca. 19,750 ± 650 years BP), and from deposits of a later age in Wyoming (Kurtén and Anderson, 1980; Van Valkenburgh et al., 1990; Berta, 1995).

MCZ 16512 is a part of the William Pringle Frost collection of fossils from the "Ashley River phosphate beds" near Charleston. The Frost collection and the Reverend Robert Wilson collection from Charleston were acquired by the Museum of Comparative Zoology and prompted Allen's (1926) report on vertebrate fossils from those collections. A great many specimens accumulated during the phosphate mining days have no specific locality data, and MCZ 16512 is no exception. However, since most of the land mining operations in the Charleston area were located along the Ashley River this specimen probably came from one of the mines on the north side of the river, where the early Pleistocene (Middle Irvingtonian) Penholoway Formation underlies the Wando Formation, as reported above in the account of the sloth remains referred to *Eremotherium* sp. The worn surfaces of the specimen suggest that it was reworked from the Penholoway into the Wando Formation and was recovered from the lag deposit at the base of that unit during phosphate mining operations.

The dark coloration and considerable degree of wear evident in USNM 299910, found on Surfside Beach, Horry County, South Carolina, indicates that this specimen has been reworked. As previously noted, *M. inexpectatus* is known from sites in Florida ca.1.7 Ma in age, the latest known date for this taxon being 0.6 Ma. However, sediments of the latter age have not been preserved in the Surfside Beach-Myrtle Beach area. Along that expanse of the coastline the Socastee Formation (early Late Pleistocene; 0.12 Ma) and the upper bed of the Waccamaw Formation (Early Pleistocene; 1.6-1.4 Ma) are the only existing Pleistocene units from which USNM 299910 might have been reworked. The Socastee is much too young to have furnished the specimen, so the Waccamaw is its most likely origin. This specimen and MCZ 16512 provide the first evidence of *Miracinonyx* from the region between Maryland (Cumberland Cave) and West Virginia (Hamilton Cave) and the Florida localities (Inglis IA and Leisey Shell Pit 3) but it would certainly not be unexpected within this region.

* * *

Genus *Puma* Jardine, 1834

PUMA CONCOLOR (LINNAEUS, 1758)

Figures 18, 24

MATERIAL.—Right mandibular ramus (DM 3) missing symphyseal region and posterior portion behind m1 but with alveolus for P3 and with p4 and incomplete m1 in place.

LOCALITY AND HORIZON. S.C., Charleston Co.; Edisto Beach; Gracie Marvin, 11 March 1991; undetermined offshore unit; Late Pleistocene.

AGE.—Rancholabrean, Late Wisconsinan.

DISCUSSION.—*Puma concolor* has the greatest range of any Recent mammal in the Western Hemisphere, occurring from British Columbia to Patagonia (Hall and Kelson, 1959:955), and is one of the most widely-distributed carnivoran species in the world (Kurtén and Anderson, 1980: 194). It appears first in the early Pleistocene Froman Ferry fauna of Idaho (c. 1.6 Ma), apparently having evolved from *Felis lacustris* of the Hagerman fauna (3.6 Ma) (Repenning et al., 1995).

As noted above, a mandibular ramus (MCZ 16512) questionably suggested as *Puma concolor* by Allen (1926:448) is referable to the cheetah-like cat *Miracinonyx*. A more recent specimen (DM 3, Figure 24) from Edisto Beach, Charleston County, appears to be the first clear evidence of cougars in the fossil record of South Carolina. The anterior portion of this specimen is missing from the posterior wall of the canine alveolus forward, and the entire region posterior to the beginning of the ascending process is also lost. Only the alveolus and posterior root of p3 is preserved, but p4 is intact and shows occlusal wear on all three cusps. The anterior cusp of m1 is well preserved and has little wear, but the posterior margin and apex of the posterior cusp have been broken off. This specimen is in the private collection of the late Don Marvin of Edisto Beach, South Carolina. Its measurements are as follows:

Greatest anteroposterior length of ramus as preserved	82.3
Depth of ramus at center of m1, labial side	26.7
Alveolar length, p3-m1	46.3
Length of alveolus, p3	12.9
Anteroposterior length of p4 at base of crown	16.4
Ventral margin of ramus to highest point of preserved portion of m1	37.5

This specimen appears to be the only Pleistocene record of this species along the Atlantic coast north of Florida, the only southeastern state that Kurtén and Anderson (1980:194) listed as having furnished fossil *P. concolor* remains.

Hall and Kelson (1959:957) show two subspecies ("*Felis*" *c. concolor* and "*Felis*" *c. coryi*) occurring in South Carolina in Recent times, and those races may well have been in this region during Late Pleistocene times, but the fossil specimen certainly offers no clue to its subspecific identity.

* * *

Genus *Lynx* Kerr, 1792
Lynx rufus (**Schreber, 1777**)
Figures 18, 25

MATERIAL.—ChM PV3462, right humerus, missing proximal end.

LOCALITY AND HORIZON.—S.C., Charleston Co.; Edisto Beach; Magaret Pulliam, summer 1980; undetermined offshore unit; Late Pleistocene.

AGE.—Rancholabrean, Late Wisconsinan.

DISCUSSION.—The bobcat has been reported from the Pleistocene of South Carolina by Hay (1923a:363) and by Bentley et. al. (1995:10). The present specimen is the first evidence of this species in the late Pleistocene mammal remains from Edisto Beach. A well-preserved right humerus missing only the head (Figure 25), it measures 126.2 mm in length and 25.3 mm in transverse diameter at the distal end. In size and morphology it matches Recent bobcat humeri in the Charleston Museum mammal collection. *L. rufus* is one of the most common members of North American Pleistocene faunas, having been recorded in fossil form from Late Blancan to Late Ranchlabrean time (Kurtén and Anderson, 1980:197). Today, bobcats are fairly well distributed over most of South Carolina but are more common in the coastal plain river swamps.

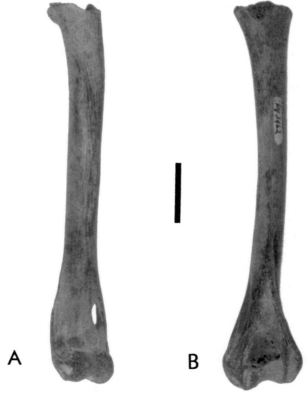

Figure 25. ChM PV3462, *Lynx rufus,* right humerus missing proximal end. **A**, anterior view; **B**, posterior view. S.C., Charleston Co., Edisto Beach, undetermined offshore unit (Late Pleistocene). Scale bar = 20 mm.

Family ODOBENIDAE
Subfamily ODOBENINAE
Genus *Odobenus* Brisson, 1762
ODOBENUS ROSMARUS (LINNAEUS, 1758)
Figures 26, 27a

MATERIAL.—AMNH 104788, proximal portion of tusk (casts, ChM PV2794, USNM 263614); AMNH 104790, virtually complete tusk (casts, ChM PV2795, USNM 26315); ChM PV5751, distal end of tusk; ChM PV5752, distal end of tusk; ChM PV5754, fragment of tusk.

LOCALITY AND HORIZON.—AMNH 104788, 104790: S.C., Charleston Co.; phosphate mines near Charleston; Wando Formation (?), Late Pleistocene. ChM PV5751: S.C., Charleston Co.; dredged from Wando River, ca. 1880; Wando Formation (?), Late Pleistocene. ChM PV5752: S.C., Charleston Co.; (?)Wando River, ca. 1880; Wando Formation (?), Late Pleistocene. ChM PV5754: S.C., Horry Co.; excavation for marina on Cedar Creek adjacent to U.S. Route 17, 1.2 miles (1.93 km) northeast of S.C. Route 9; C.B. Berry, 1977; Late Pleistocene (formation undetermined).

AGE.—AMNH 104788, 104790; ChM PV5751, PV5752, PV5754; Rancholabrean, Sangamonian.

DISCUSSION.—Fossil walrus remains were first reported from South Carolina by Joseph Leidy (1877:214-216, Pl 30, fig.6), who gave notice of a tusk in a collection of fossils from the Ashley River phosphate beds exhibited by the Pacific Guano Company of Charleston at the United States Centennial Exposition in Philadelphia in 1876. He was concerned about the future of those specimens, "for in most cases they are likely to be lost to the inspection of students after the great exhibition closes" (Leidy, 1877:209). Evidently, his attempts to obtain some of the specimens for the Academy of Natural Sciences of Philadelphia were unsuccessful, for in a footnote to the remarks above he observed that "The finder and unscientific owner of fossils, ignorant of their real importance, often retain them as curiosities, with exaggerated notions of their pecuniary value, and no argument is sufficient to induce them to part with the specimens or place them where they may be accessible to the student" (Leidy, 1877:209). His fears about the collection were not unfounded, because the present whereabouts of the walrus tusk and the other specimens that he reported and figured (Leidy, 1877) are unknown.

O.P. Hay (1923a:363) included *O. rosmarus* in his list of Pleistocene mammals from South Carolina, almost certainly based upon two tusks that he examined at The Charleston Museum in 1915 and recorded in the notes that he made of the specimens that he saw during his visit to Charleston. One specimen (ChM PV5751) was recovered from the dredging of phosphate on the bottom of the Wando River near Charleston during the nineteenth century, and the other (ChM PV5752) also was found during mining operations, perhaps dredged from the bottom of the Wando River.

Ray et al. (1968:16) made note of cranial fragments tentatively referred to *O. rosmarus* from Carolina Beach, Horry County, South Carolina, and from Edisto Island.

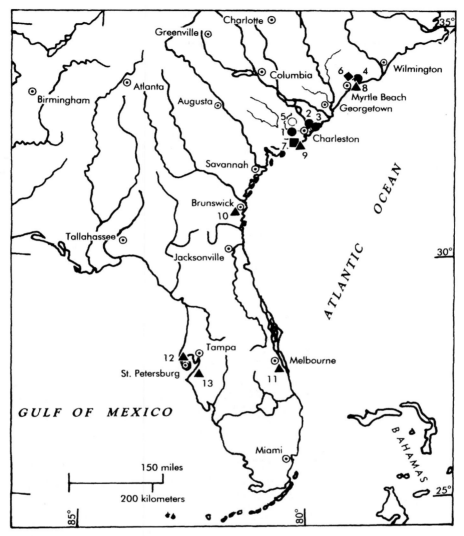

Figure 26. Pleistocene localities for *Odobenus rosmarus* (● 1-4), *Odobenus* sp. (○ 5), *Erignathus barbatus* (◆ 6), and *Pseudorca crassidens* (■ 7) from South Carolina, and *Monachus tropicalis* (▲ 8-13) from the southeastern United States. 1, AMNH 104788, 104790, Charleston Co., phosphate mines near Charleston, Wando Fm. (Late Pleistocene); 2, ChM PV5751, Charleston Co., Wando River, Wando Fm. (Late Pleistocene); 3, ChM PV5752, Charleston Co., (?)Wando River, Wando Fm. (Late Pleistocene); 4, ChM PV5754, Horry Co., excavation for marina on Cedar Creek; undetermined late Pleistocene unit. 5, ChM PV5753, Dorchester Co., ditch on Trolley Road, Ten Mile Hill Beds (Late Middle Pleistocene). 6, Horry Co., Intracoastal Waterway at Myrtle Beach, Socastee Fm. (Late Pleistocene). 7, ChM PV5023, Charleston Co., Edisto Beach, undetermined offshore unit (Late Pleistocene). 8, USNM 425911, Horry Co., Intracoastal Waterway near Crescent Beach, Socastee Fm. (Late Pleistocene); 9, DM 5 (cast, ChM PV5717), Charleston Co., Edisto Beach, undetermined offshore unit (Late Pleistocene); 10, CK 1 (cast, ChM PV5721), Georgia, Glynn Co., Brunswick (Late Pleistocene); 11, Florida, Brevard Co., Melbourne (Late Pleistocene) (Ray, 1958); 12, Florida, Pinellas Co., St. Petersburg (Late Pleistocene) (Ray, 1961); 13, Florida, Hillsborough Co., Leisey shell Pit 1A (Early Pleistocene) (Berta, 1995).

Roth and Laerm (1980:16) also noted the cranial fragment from Edisto Island and indicated that it is in the collection at Hampden-Sydney College.

A large, virtually complete walrus tusk (AMNH 104788) and the proximal end of a tusk (AMNH 104790) from the Ashley River phosphate beds were in the Cohen collection of fossils when it was received by the American Museum of Natural History in 1908. Casts of those specimens are at the National Museum of Natural History (USNM 26314, 26315) and The Charleston Museum (ChM PV2794, PV2795).

The heavily mineralized middle third of a walrus tusk (ChM PV5754, Figure 27a) was discovered in 1977 by Mr. C.B. Berry during excavations for the marina at Cedar Creek on U.S. Route 17 1.93 km northeast of South Carolina Route 9 in Horry County. As preserved, it is 149 mm in proximodistal length, 78 mm in greatest anteroposterior diameter, and 48 mm in greatest transverse diameter. The specimen was collected in a heavily fossiliferous shell bed near the top of the section as seen in photographs made by Mr. Berry at the time of the excavation. He also collected a large specimen of the bivalve mollusk *Mercenaria campechiensis*. Now inundated by the waters of the marina, the section is no longer available for inspection. Richard E. Petit examined the photographs and recognized the deposits as a late Pleistocene

Figure 27. Fragments of walrus tusks from South Carolina. **A**, ChM PV5754, *Odobenus rosmarus*, Horry Co., marina excavation on Cedar Creek; undetermined late Pleistocene unit. **B**, ChM PV5753, *Odobenus* sp., distal end of tusk, Dorchester Co., ditch on Trolley Road (County Road 199); Ten Mile Hill Beds (Late Middle Pleistocene). Scale bars = 30 mm.

unit, but the specific formation was not readily distinguishable to him. The late Pleistocene Socastee Formation is the most likely source of this specimen.

A total of five specimens of *Odobenus rosmarus* from the late Pleistocene of South Carolina are reported above, and those, coupled with the two records reported by Ray et al. (1968:16), intimate that walruses were more than casual visitors along the coast of South Carolina during Pleistocene time. Ray et al. (1968:16) have documented Pleistocene records of this species from the coasts of North Carolina and Virginia, noting that "These specimens as well as many others from the Atlantic coast are not specifically identifiable on strictly morphological grounds, but are referred to *O. rosmarus* on the premises that Atlantic and Pacific walruses are conspecific, and that the fossils are indistinguishable from corresponding elements of the living species." The South Carolina specimens reported above are referred to *O. rosmarus* on the same grounds, isolated elements of Late Pleistocene walruses from the Atlantic coast apparently representing that species. The additional specimens from South Carolina suggest that the natural range of the walrus extended even further south to the coast of the Carolinas during Pleistocene time, the reduction to its present range perhaps having been induced by pressure from increasing numbers of Paleo-Indians occupying the coastal regions. As Ray et al. (1968:21) have commented, "the walrus bred on Sable Island within historic time and has receded steadily northward from that point and continues to do so. There is no reason to suppose that this restriction is climatically controlled, and much reason to attribute it directly to the activities of Man. ... climate, notably temperature, unquestionably is a potential limiting factor in (seasonal) distribution ... and probably was the critical factor prior to the ascendancy of Man. However, the modern distribution of walruses undoubtedly represents only that fraction of their former wider niche least frequented by their most relentless enemy, Man."

Multiple records of walruses on the coast of South Carolina, most of them probably from the Sangamonian interglacial period, would seem to confirm those observations and also to infer that walruses of Pleistocene times had a higher tolerance for temperate climates than might be supposed from the present distribution of the species in the northern polar regions.

* * *

ODOBENUS SP.
Figures 26, 27b

MATERIAL.—ChM PV5753, fragments of tusk.

LOCALITY AND HORIZON.—S.C., Dorchester Co.; ditch on New Trolley Road (County Road 199); Matthew Swilp, 21 August 1983; Ten Mile Hill Beds, Late Middle Pleistocene.

AGE.—Rancholabrean; Mid Illinoian.

DISCUSSION.—These fragments comprising the greater portion of a small tusk (ChM PV5753, Figure 27b) are referred to *Odobenus* because of general resemblances to tusks of *O. rosmarus* discussed above. If that assignment is correct, ChM PV5753 represents the earliest known evidence of *Odobenus* in the fossil record of

South Carolina. The two largest elements are the distal end (119 mm in length) and a larger piece (120 mm in length) that appears to be the proximal end or very close to it. There are other small fragments, but the pieces that would unite the two larger portions are missing. The specimen is much smaller than any of the tusks of *O. rosmarus* reported above, being only 56 mm at its greatest anteroposterior width and 25 mm in greatest transverse diameter. The original length of the tusk cannot be determined absolutely, but it seems to have been approximately 226 mm long proximodistally, based on the ratio of anteroposterior width (82 mm) to proximodistal length (330 mm) obtained from a cast (ChM PV2795) of AMNH 104790, a virtually complete tusk of *O. rosmarus* mentioned above. In any event, PV5753 is certainly the tusk of a small individual, probably a young animal. As is typical of walrus tusks, the anterodistal surface of the specimen is noticeably worn.

This specimen has not been referred to *Odobenus rosmarus* partly because of its size and partly because of the uncertainty regarding the presence of that species along the coast of South Carolina during Late Middle Pleistocene time. Whether it represents a presently unknown species of walrus smaller than *O. rosmarus* or merely the tusk of a young individual of the latter species cannot be answered without additional material.

Whatever the case with regard to its specific identity, PV5753 is of particular interest because of its zoogeographic implications. As mentioned earlier in this paper, the Ten Mile Hill Beds have been dated at 0.24-.020 Ma (Weems and Lemon, 1984a), which places this unit within the mid-Illinoian interglacial period that separated the two periods of Illinoian glaciation (Richmond and Fullerton, 1986:8). Again, as with the *O. rosmarus* specimens discussed above, we see apparent evidence of walruses along the coast of South Carolina during an interglacial period, reiterating the implication that these pinnipeds have had a broader climatic tolerance than is indicated by the distribution of the living species.

* * *

Family PHOCIDAE
Subfamily PHOCINAE
Genus *Erignathus* Gill, 1866
ERIGNATHUS BARBATUS (ERXLEBEN, 1777)
Figures 26, 28

MATERIAL.—Distal end of left humerus (ChM PV5399).

LOCALITY AND HORIZON.— S.C., Horry Co.; south bank of Intracoastal Waterway behind Shriner's Club building on north side of U.S. Rt. 17, 0.8 km southwest of merger with U.S. Rt. 17 Business Route through Myrtle Beach; Robert L. Johnson, April 1995. Socastee Formation, Late Pleistocene.

AGE.—Rancholabrean, Sangamonian.

DISCUSSION.—Although Late Oligocene beds in the vicinity of Charleston have furnished the oldest known pinniped remains (Koretsky and Sanders, in press), fossil representatives of that group are relatively scarce in South Carolina. Previously published records of pinnipeds in the Pleistocene of South Carolina consist only of the report by Ray et al. (1968:16–19) of a partial right innominate bone (ChM 51.23.1) from Edisto Beach tentatively referred to *Halichoerus grypus* Fabricius, the Gray Seal, and cranial fragments from Carolina Beach and Edisto Beach referred with less certainty to the Atlantic Walrus, *Odobenus rosmarus* (Linnaeus).

Additional material has been accumulating in the Charleston Museum collection. In 1959 the Museum received a collection of Pleistocene vertebrate fossils collected by the late Ivan R. Tompkins of Savannah, Georgia. Among those specimens was the right second proximal phalanx of a pinniped (ChM PV5796)—more than likely a phocid—collected by Tompkins near Savannah in November 1937. This specimen is 64.5 mm in proximodistal length and 14.25 mm in transverse diameter at the distal end and apparently does not represent a very large individual. A newly acquired specimen (ChM PV5040), the acetabular region of a right innominate bone of a pinniped, has been identified as that of a phocid seal of undetermined specific or generic allocation (Irina Koretsky, personal communication, November 1994). Collected by Dale Theiling on the bottom of the Toogoodoo River in Charleston County in 1991, this specimen is presumed to be of Pleistocene age on the basis of other mammalian elements from the same locality.

More recently, the distal end of a left humerus of a pinniped was recognized among several vertebrate fossils collected by Robert L. Johnson of Myrtle Beach, South Carolina, and brought to The Charleston Museum for identification. Comparison with material in the U.S. National Museum of Natural History demonstrated that this specimen (ChM PV5399, Figure 28) is referable to the Bearded Seal, *Erignathus barbatus* (Irina Koretsky, personal communication, April 1996). The subspecies *E. b. barbatus* (Erxleben) occurs today along the Arctic shores of northern Europe and North America from Greenland to the Bering Sea and Hudson Bay and on the North Atlantic coast southward to Labrador and, rarely, to Newfoundland (Anderson, 1946:80). An Alaskan form, *E. b. nauticus* (Pallas) ranges from Point Barrow southward along the coast of Alaska to Bristol Bay (Hall and Kelson, 1959:983). "This large seal, the largest Phocid of the northern seas, appears to be nowhere abundant, and is usually described as rather solitary, avoiding the company of other species, and as never occurring in large herds like the Harp Seal" (Allen, 1880:670).

Previous evidence of *E. barbatus* in the fossil record is sparse, consisting of only five Pleistocene records, three of which are from North America and two from Europe (Ray et al., 1982). These are: a partial skull and "probably associated" atlas from deposits near Finch, Ontario, Canada (Harington, 1977:521), probably 10,500–11,800 years in age (Ray and Spiess 1981:426); three caudal vertebrae and 21 bones from the hind limbs, "apparently from a single individual" found near Orrington, Maine, in a blue clay approximately 12,200–12,600 years old (Ray and Spiess 1981:423–426); "«fossil remains» of unspecified nature from «Post-tertiary Beds»" on Ellesmere Island, Canada (Ray et al., 1982:2–3), originally reported by Feilden (1877:488); two limb elements from near Overstrand, Norfolk, England, supposedly collected from the «Forest Bed» (Newton, 1889:147-148; 1891:19) in the 19th century and possibly of Early Pleistocene age (Ray et al., 1982:3); and miscellaneous skeletal elements, a partial skeleton, and a nearly complete skeleton from

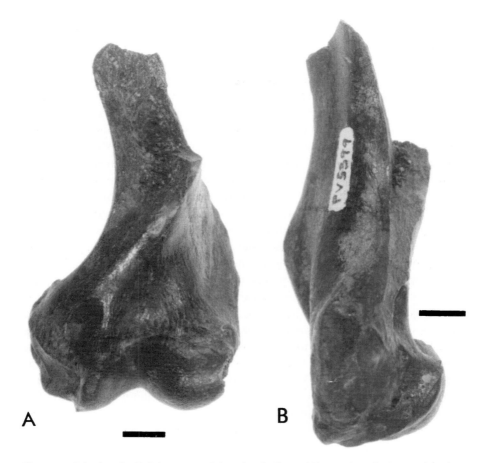

Figure 28. Distal end of left humerus, *Erignathus barbatus* (ChM PV5399). **A**, anterior view, **B**, medial view. S.C., Horry Co., Socastee Fm. (Late Pleistocene). Scale bars = 10 mm.

seven localities in the vicinity of Lake Vänern and Göteborg in southwestern Sweden (Fredén, 1975:12, 37–39; Ray et al., 1982:3).

The South Carolina specimen (ChM PV5399) is therefore of considerable significance in that it appears to be just the fourth known occurrence of *E. barbatus* in the fossil record of North America, the remains from Orrington, Maine, (Ray and Spiess, 1981) being the only previous record from the United States.

ChM PV5399 is the distal end of a left humerus broken off from the proximal portion at the narrowest region of the shaft just below the deltoid crest. Measurements (in mm) are as follows:

Proximodistal length, as preserved	90.2
Greatest transverse diameter	51.0
(axis through ectepicondyle and entepicondyle)	

The specimen was found in a lag deposit of fluvial sands and pebbles unconformably overlying the Waccamaw Formation (Early Pleistocene) on the south bank of the Intracoastal Waterway behind the Shriner's Club building on the north side of U.S. Route 17 Bypass near the northeastern city limits of Myrtle Beach, 0.8 km southwest of the northern terminus of the U.S. Route 17 Business Route through Myrtle Beach, Horry County, South Carolina (33° 46′ N, 78° 49′ W, USGS Nixonville 15′ quadrangle). In this region, the late Pleistocene Socastee Formation overlies the Waccamaw Formation (Owens, 1989), and at the seal locality it is approximately 1.1 to 1.2 m thick. The fluvial deposit at the base consists of very coarse sub-rounded grains of quartz and some feldspar, and there are numerous flat or rounded pebbles, some reaching a diameter of 22.5 mm, a specimen of that size being present in the matrix retained with the humerus (ChM PV5399). The collector, Robert Johnson, pays frequent visits to the Intracoastal Waterway in search of fossils and has observed that the coarse sand deposit is of intermittent occurrence along the Waterway. This deposit has also furnished other vertebrate remains, including two fragments of molars of *Mammuthus* sp., an incisor of *Castoroides*, a large crocodilian scute, and the crown of a mosasaur tooth, all of which are in Mr. Johnson's private collection. The mosasaur tooth is of late Cretaceous age, and Bruce R. Erickson (Science Museum of Minnesota), who determined both the scute and the tooth, suggests that the two are probably of the same age (Erickson, personal communication, October 1995). Thus, the fossils occurring in the lag deposit are of a mixed lot; some terrestrial mammals (*Mammuthus* and *Castoroides*) and a marine-mammal (*Erignathus barbatus*) from the Pleistocene and a crocodile and a marine reptile from the Cretaceous, indicating that the stream that laid down the lag deposit in Socastee time was cutting through beds of more than one age. The Cretaceous specimens were almost certainly derived from the Upper Cretaceous Pee Dee Formation, which underlies the Waccamaw Formation in that region and is exposed at certain places along the Intracoastal Waterway near Myrtle Beach.

With the exception of some smoothing of the broken edges, the *E. barbatus* humerus is not badly worn, indicating that it had not been transported for any great distance and is not reworked from the Waccamaw, which in turn suggests that the animal was in the area during deposition of the Socastee Formation in Sangamonian time approximately 120, 000 ybp (R.E. Weems, personal communication, July 2001).

The presence of *E. barbatus* so far south of its known present range and more than 10 degrees of latitude south of its southernmost Pleistocene record in Maine (44° 42′ N [Ray et al., 1982:2]) is surprising, but there are records of Recent individuals that have strayed southward to the British Isles and the coasts of France and Spain, and one as far south as the estuary of the Mondego River on the coast of Portugal at 40° 09′ N (Ray et al.,1982:8–10). Thus, the Myrtle Beach specimen, found less than seven degrees (33° 46′ N) south of the latter record, is perhaps not so remarkable as it might seem. But given the current range of this species, it would certainly not be expected in the Pleistocene of South Carolina and may even be a unique occurrence.

Subfamily MONACHINAE
Genus M*onachus* Fleming, 1822
MONACHUS TROPICALIS (GRAY, 1850)
Figures 26, 29, 30, 31

MATERIAL.—USNM 425911, right mandibular ramus with p3 and p4 but miss-ing anterior and posterior ends (cast, ChM PV5719); DM 5, right mandibular ramus with alveoli only and lacking anterior and posterior ends (cast, ChM PV5717); CK 1, atlas vertebra missing the transverse processes (cast, ChM PV5718).

LOCALITY AND HORIZON.—USNM 425911: S.C., Horry Co., Intracoastal Waterway at Possum Trot Golf Course near Crescent Beach, Robert L. Johnson, c. 1994; Socastee Formation, Late Pleistocene. DM 5: S.C., Charleston Co., Edisto Beach, Don and Gracie Marvin, c. 1993; undetermined offshore unit, Late Pleis-tocene. CK 1: Georgia, Glynn Co., spoil pile on Andrews Island across East River from Brunswick, Chet Kirby, summer 1993; formation unknown, Late Pleistocene.

AGE.—Rancholabrean, Sangamonian (USNM 425911); Rancholabrean, Late Wisconsinan (DM 5); Rancholabrean, Late Wisconsinan (?) (CK 1).

DISCUSSION.—Occurring within Recent times from islands off the coast of Honduras and Yucatan through the West Indies to the Bahamas and Florida, the West Indian Monk Seal, *Monachus tropicalis* Gray, was observed by Columbus on his second voyage to America in 1494 and was the first animal described from the New World. Subsequently hunted extensively for hides and meat, this species was on the verge of extinction by 1885 (Walker, 1975:1309). As Ray (1958:441) has noted, "The monk seal was so quickly decimated at the hands of modern man that its pre-Columbian range cannot be determined." Its status was certainly not improved by the collecting party led by Henry L. Ward and Fernando Ferrari-Perez, which killed 49 monk seals on The Triangle keys off the coast of Yucatan in December 1886. From those animals, only 34 skins, seven skeletons, and several skulls were obtained, and it was upon some of this material that J.A. Allen based his valuable monograph on this species (Allen, 1887:1–2). Gunter (1947) reported sightings of monk seals on the coast of Texas during the 1920s and 1930s, and "The last record of this species seems to be its occurrence in Jamaican waters in 1952" (Walker, 1975:1309).

The first fossil evidence of *M. tropicalis* from the mainland of the United States was a proximal phalanx of the right hallux of *Monachus tropicalis* (MCZ 4439) re-ported by Ray (1958) from Pleistocene deposits near Melbourne, Brevard County, Florida. Subsequently, Ray (1961:113) recorded a left maxilla of a monk seal from Quaternary dredgings at St. Petersburg, Pinellas County, Florida, clearly establishing the presence of *M. tropicalis* along the Gulf Coast of the Florida peninsula. More re-cently, Berta (1995) recorded a postcanine tooth (UF 81734) of this species from early Pleistocene deposits in Leisey Shell Pit 1A in Hillsborough County, Florida. Until now, those specimens have represented the northernmost records of *M. tropi-calis*.

During the summer of 1993, Chet Kirby of Brunswick, Glynn County, Geor-gia, found a well preserved mammalian atlas vertebra (Figure 29) on a pile of

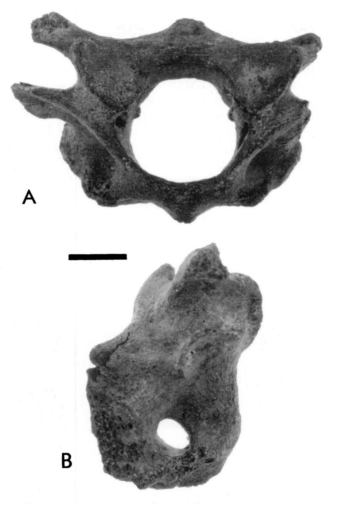

Figure 29. Atlas vertebra, *Monachus tropicalis* (CK 1). **A,** anterior view; **B,** left lateral view. Georgia, Glynn Co.; Brunswick, spoil pile on Andrews Island; undertermined late Pleistocene unit. Scale bar = 10 mm.

dredgings from the Turtle River on Andrews Island across the East River from the city of Brunswick (USGS Bruswick East and Brunswick West 15′ quadrangles) (Figure 26). This specimen was identified as *Monachus tropicalis* by Clayton E. Ray (U.S. National Museum of Natural History). It is missing the outer portions of both transverse processes but is well preserved otherwise. As a dredge pile specimen, its stratigraphic origin cannot be determined with certainty, but it is not dark and heavily permineralized as are most of the Pliocene specimens from that area (Richard Hulbert, private communication, September 1999) and thus is assumed to have come from the same as-yet-undetermined late Pleistocene unit that furnished several vertebrate elements of Rancholabrean age reported by Hulbert and Pratt (1998, table 1).

Measurements (in mm) of the specimen are as follows:

Greatest anteroposterior length	43.7
Greatest transverse diameter (as preserved)	82.5
Greatest vertical diameter (as preserved)	58.7
Greatest transverse diameter, anterior face of centrum	66.5
Greatest transverse diameter, posterior face of centrum	57.5
Transverse diameter, neural canal	29.2
Vertical diameter, neural canal	32.7

The collector has retained possession of the specimen, but casts of it are in The Charleston Museum (ChM PV5721) and the National Museum of Natural History.

Two additional specimens identified by Clayton E. Ray document the presence of *M. tropicalis* even further north along the Atlantic coast during Late Pleistocene time. A

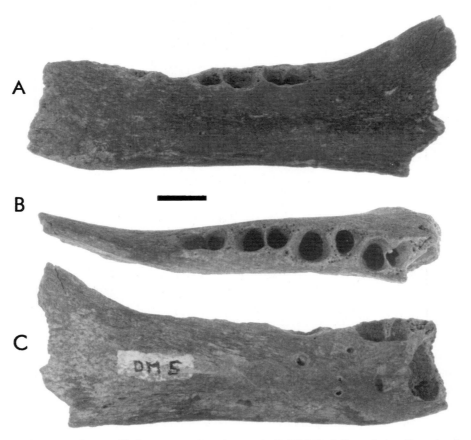

Figure 30. Right mandibular ramus, *Monachus tropicalis* (DM 5). **A**, lingual view; **B**, occlusal view; **C**, labial view. S.C., Charleston Co., Edisto Beach; undetermined offshore unit (Late Pleistocene). Scale bar = 10 mm.

right mandibular ramus (DM 5, Figure 30) with alveolae for p2, p3, p4, and m1, found on Edisto Beach, Charleston County, South Carolina by Don and Gracie Marvin, is extremely small and obviously belonged to a very young individual. Missing the anterior end beyond the alveolus for p2 and the posterior end behind the foot of the ascending process, the specimen measures 83 mm in anteroposterior length as preserved.

Another partial right mandibular ramus (USNM 425911, Figure 31) was found

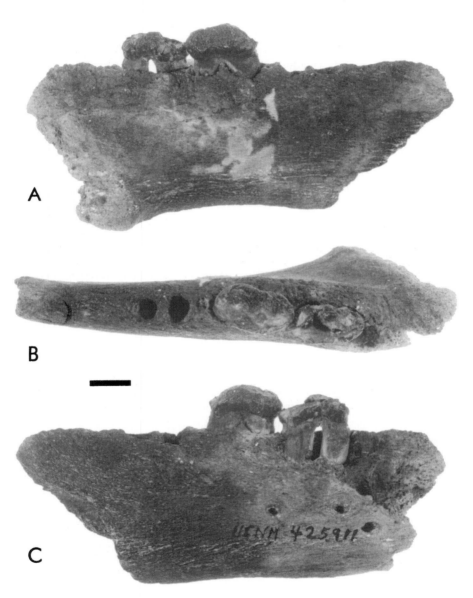

Figure 31. Right mandibular ramus, *Monachus tropicalis* (USNM 425911) with p3 and p4 in place. **A**, lingual view; **B**, occlusal view; **C**, labial view. S.C., Horry Co., Socastee Fm. (Late Pleistocene). Scale bar = 10 mm.

by Robert L. Johnson along the Intracoastal Waterway at the Possum Trot Golf Course near the town of Crescent Beach, Horry County, South Carolina (USGS Wampee 7.5′ quadrangle), the golf course being located on a secondary road off U.S. Route 17, 3.9 miles (6.27 km) southwest of S.C. Route 65. This specimen is that of a considerably larger individual than the Edisto Beach specimen (DM 5) and has the third and fourth postcanine in place.

Measurements (in mm) of the South Carolina specimens are as follows:

	USNM 425911	DM 5
Anteroposterior length of ramus as preserved	99	83
Anterior margin of alveolus for p2 to posterior margin of alveolus for m1	64	48
Depth of ramus at m1	36	22

In his table comparing osteological measurements of Recent specimens of *M. tropicalis* and four other phocid taxa, Allen (1887:9) gives the depth of the mandibular ramus at the "last molar" (m1) as 34 mm in an adult male, 28 mm in an adult female, and 21 mm in a specimen that Allen (1887:2) identified as a "suckling young one." Allen's (1887:9) measurements at that point show the mandibular ramus of *M. tropicalis* to be considerably more robust than in *Erignathus barbatus* (23 mm), *Phoca vitulina* (21 mm, 23 mm), and *Cystophora cristata* (31 mm), the latter taxon most closely approaching *M. tropicalis* in depth. Measuring 36 mm in depth at the last postcanine, USNM 425911 from Horry County, South Carolina, is clearly an adult. The Edisto Island specimen (DM 5) is 22 mm in depth at that point, virtually the same as that of Allen's (1887:2, 9) youngest individual (21 mm), indicating that it represents a nursing pup. If such is the case, this specimen would seem to document a breeding population of *M. tropicalis* as far north as the coast of South Carolina during Rancholabrean time.

The specimens reported herein add to the fossil record of this species and extend its known range during the Pleistocene from Florida (Ray, 1958, 1961) northward to Georgia and thence further northward along the coast to Charleston County and Horry County, South Carolina.

* * *

Order CETACEA
Family DELPHINIDAE
Genus *Pseudorca* Reinhardt, 1862
PSEUDORCA CRASSIDENS (OWEN, 1846)
Figures 26, 32

MATERIAL.—ChM PV5023, right periotic.
LOCALITY AND HORIZON.—S.C., Charleston Co.; Edisto Beach; Doris Holt, 21 August 1983; undetermined offshore unit, latest Pleistocene-early Holocene(?).
AGE.—Latest Wisconsinan-early Holocene(?).

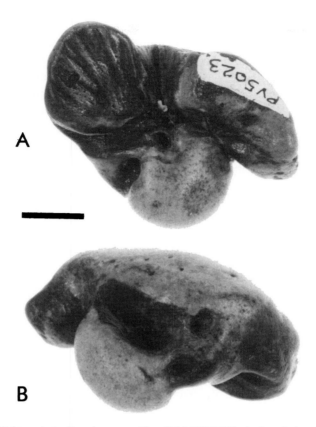

Figure 32. Right periotic, *Pseudorca crassidens* (ChM PV5023). **A**, dorsal view; **B**, ventral view. S.C., Charleston Co., Edisto Beach; undetermined offshore unit (Late Pleistocene). Scale bar = 10 mm.

DISCUSSION.—The type of this species is a subfossil specimen from the Lincolnshire fens, England, that was lost during the nineteenth century (True, 1889: 143). A pelagic form, the false killer whale occurs today in tropical to warm temperate waters of the western North Atlantic, having been reported from off Maryland and the east coast of North America, in the Gulf of Mexico, and southward through the Caribbean Sea to Venezuela (Leatherwood et al., 1976). Elsewhere, it is distributed through the tropical and warm temperate Pacific and Indian Oceans (Evans, 1987:55). Adults reach a length of at least 5.5 m (Evans, 1987:55).

ChM PV5023, a right periotic somewhat worn but otherwise in good condition, is the first evidence of *P. crassidens* in the fossil record of South Carolina. The specimen was collected by Doris Holt on Edisto Beach, Charleston County, on 21 August 1983. It is 45.3 mm in anteroposterior length (posterior end of posterior process to anterior margin of anterior process) and 31.4 mm in transverse diameter (plane of lateral margins of anterior and posterior processes to medial margin of promontorium). Records kept for many years at the Charleston Museum reflect no known strandings of this species on the South Carolina coast within historic times, despite its occurrence in pelagic waters offshore.

A skull of *Pseudorca* sp. from the early Pliocene Yorktown Formation at the Lee Creek mine in North Carolina has been reported by Whitmore (1991:224). I have not attempted to search out other fossil occurrences of *Pseudorca*, the principal purpose of this report being merely to place the Edisto Beach specimen on record. As noted elsewhere in this paper, the marine-mammal specimens from Edisto Island seem almost certainly to be of latest Wisconsinan or early Holocene age.

* * *

Order RODENTIA
Infraorder CASTORIMORPHA
Family CASTORIDAE
Subfamily CASTORINAE
Tribe Castorini
Genus *Castor* Linnaeus, 1758
Castor canadensis Kuhl, 1820
Figures 33, 34

MATERIAL.—ChM PV6999, right mandibular ramus with p4, mi, m2, m3, and part of i1; missing anterior portion of i1, ventral portion of mandible along alveolus for i1, and coronoid region.

LOCALITY AND HORIZON.—S.C., Charleston Co.; Leadenwah Creek, Wadmalaw Island (N. 32° 39.6′, W. 80° 39.6′); Barry Albright, summer 1986; Wando Formation, Late Pleistocene.

AGE.—Rancholabrean, Sangamonian,

DISCUSSION—Pleistocene remains of the common beaver have been reported from South Carolina by Leidy (1860), Hay (1923a), Roth and Laerm (1980), and Bentley et al. (1995) and are abundantly represented across North America. The earliest known record of *C. canadensis* is from the late Blancan Haile 15 site in Florida (Kurtén and Anderson, 1980).

Leidy (1860, pl. 21, fig. 2) noted and figured three molar teeth of *Castor* from the vicinity of the Ashley River near Charleston, and Hay (1923a) also listed a specimen from along the Ashley River but did not specify the element and did not indicate whether he was citing Leidy's (1860) record or reporting another one. Roth and Laerm (1980) recorded a right femur (HS 104) from Edisto Beach in the Hampden-Sydney College collection but suggested that it might represent a modern individual. That possibility seems unlikely, however, because beavers inhabit freshwater streams and not beaches. Suitable beaver habitat has not been present in that area since the late Pleistocene. Bentley et al. (1995) documented *C. canadensis* in their late Pleistocene Ardis local fauna from Dorchester County on the basis of three teeth and six postcranial elements.

In 1986 Barry Albright collected an incomplete but well-preserved right mandibular ramus (ChM PV6999, Figure 34) in the Wando Formation on the bank of Leadenwah Creek on Wadmalaw Island, Charleston County. Athough i1 is preserved only as a fragment in the symphysial region, all of the molariform dentition is in place and displays the enamel folding typical of *C. canadensis*. The ventral region of the ramus, including the remainder of i1, is missing, as well as the entire posterior end of the specimen. The degree of breakage permits only two measurements (in mm) of ChM PV6999:

Figure 33. Locality records for *Neofiber diluvianus* (● 1), *Neofiber* cf. *N. diluvianus* (● 2), and *Neofiber alleni* (○ 3-4) from the Pleistocene of eastern North America, and *Castor canadensis* (□ 5) and *Erethizon dorsatum* (■ 6) from South Carolina. **1,** Type locality: Pennsylvania, Montgomery Co.; Port Kennedy Cave (Middle Pleistocene) (Cope, 1896; Hibbard, 1955). **2,** ChM PV5070, PV5398, N.C., Brunswick Co.; Calabash; Waccamaw Fm. (Early Pleistocene). **3,** ChM GPV2018, S.C., Charleston Co., ca. 19 km N. of Charleston; Wando Fm. (Late Pleistocene). **4,** ChM PV5071, S.C., Dorchester Co., ca. 33.7 km NW of Charleston; Ten Mile Hill Beds (Late Middle Pleistocene). **5,** ChM PV6999, S.C., Charleston Co.; Leadenwah Creek, Wadmalaw Island; Wando Fm. (Late Pleistocene). **6,** DM 1, S.C., Charleston Co.; Edisto Beach; undetermined offshore beds (Late Pleistocene).

Total anteroposterior length (as preserved)	88.3
Anteroposterior length of tooth row	31.2

Escept for its missing portions, the specimen is in excellent condition and appears to be the first mandibular ramus of *C. canadensis* reported from the Pleistocene of South Carolina.

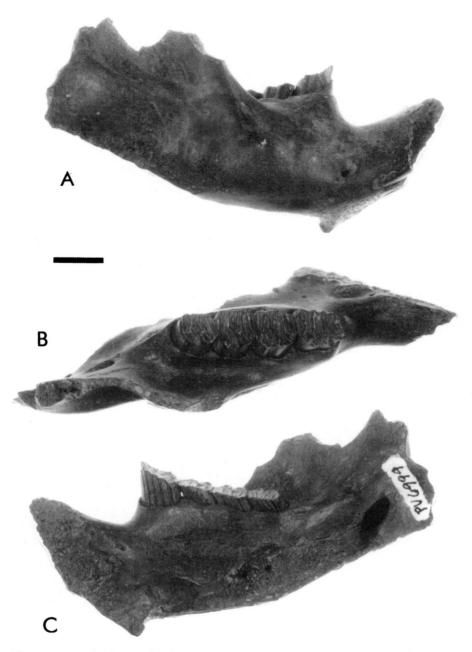

Figure 34. Partial right mandibular ramus, *Castor canadensis* (ChM PV6999) with fragment of i1 and p4-m3. **A**, labial view; **B**, occlusal view; **C**, lingual view. S.C., Charleston Co., Wando Fm. (Late Pleistocene). Scale bar = 10 mm.

Suborder MYOMORPHA
Family CRICETIDAE
Subfamily ARVICOLINAE
Genus *Neofiber* True, 1884
NEOFIBER ALLENI TRUE, 1884
Figures 33, 35, 37

MATERIAL.—ChM GPV2018, partial right mandibular ramus with root of incisor, m1, and m2 in place but missing coronoid region; ChM PV5071, left m2 missing posterior margin.

LOCALITY AND HORIZON.—ChM GPV2018: S.C., Charleston Co.; Sears Roebuck construction site, Northwoods Mall shopping center, ca. 19 km N. of Charleston; Doris Holt, August 1972; Wando Formation, Late Pleistocene. ChM PV5071: S.C., Dorchester Co.; ditch on Trolley Road (County Road 199) adjacent to Baptist Mission, ca. 21 mi. (33.7 km) northwest of Charleston; Matthew Swilp, December 1983; Ten Mile Hill Beds, Late Middle Pleistocene.

AGE.—GPV2018, Rancholabrean, Sangamonian; PV5071, Irvingtonian, Illinoian.

DISCUSSION.—In August 1972, excavations for the construction of a shopping center north of Charleston, Charleston County, South Carolina, produced a fossilized rodent mandible that provides another record of the Round-tailed Muskrat, *Neofiber alleni* True, in South Carolina. This species is now restricted to Florida and extreme southeastern Georgia (Hall and Kelson, 1959) but has been recorded from late Pleistocene fissure fillings in northwestern Georgia (Ray, 1967), from Sangamonian (Late Pleistocene) deposits near Savannah, Georgia (Hulbert and Pratt, 1998), and from Wisconsinan-age sediments in Dorchester County, South Carolina (Bentley et al., 1995:22). The latter record is based upon a right M1 (SCSM 93.105.153) and an incomplete molar (SCSM 93.105.154; not figured) of indeterminate position from the Late Pleistocene Ardis local fauna found at the Giant Portland Cement Company

TABLE 6. Measurements of mandibular rami and teeth of *Neofiber alleni* (ChM PV2018) from Wando Fm. (Late Pleistocene) near Charleston, S.C., and *Neofiber* cf. *N. diluvianus* (ChM PV5070), PV5398) from Waccamaw Fm. (Early Pleistocene) at Calabash, N.C., and holotype right m1 of *Schistodelta sulcata* (=*Neofiber diluvianus* [Cope]) (ANSP 140) from Port Kennedy Cave, Pa. (Middle Pleistocene). Dashes (—) indicate measurements not available.

	PV2018	PV5070	PV5398	ANSP 140
Greatest anteroposterior length of specimen, as preserved (including incisor)	23.50	29.25	25.25	- - -
Greatest depth (occlusal surface of m1 to to ventral surface of ramus)	12.60	11.70	10.58	- - -
Greatest transverse diameter of ramus	5.70	4.70	4.48	- - -
Greatest depth of ramus at m1, lingual side	8.30	7.75	6.55	- - -
Greatest depth of ramus at m2, lingual side	7.00	5.50	5.15	- - -
Anteroposterior length of tooth row, m1 and m2	8.33	7.75	7.60	- - -
Anteroposterior length, m1	5.20	4.80	4.70	4.0^{+}
Greatest anteroposterior diameter, 3rd reentrant angle, m1	0.73	0.60	0.58	0.65

quarry near Harleyville (Bentley et al., 1995). *N. alleni* is not known to have occurred north of its present range since the Pleistocene.

ChM GPV2018 is a partial right mandibular ramus with m1 and m2 in place (Figures 35, 37). The symphyseal region and the posterior portion behind the posteriormost extent of the alveolus for the incisor are missing. The specimen was found by Charleston Museum volunteer Doris Holt in spoil material from excavations for the construction of the Sears Roebuck section of Northwoods Mall shopping center,

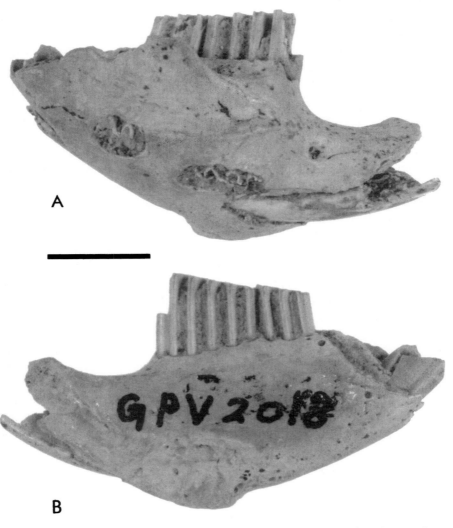

Figure 35. Partial right mandibular ramus, *Neofiber alleni* (ChM GPV2018), with m1 and m2 in place. **A**, labial view; **B**, lingual view. S.C., Charleston Co.; Northwoods Mall construction site, ca. 19 km N. of Charleston; Wando Fm. (Late Pleistocene). Scale bar = 5 mm. For occlusal view see Figure 37.

ca. 19 km north of Charleston at the intersection of U.S. Route 52 and the Route 52 exit from Interstate Highway 26 (Charleston County, USGS Ladson 7.5′ quadrangle, 32° 56.8′ N., 80° 02.6′ W.). In the absence of comparative material in our collection the mandible was submitted to Clayton E. Ray (National Museum of Natural History), who determined it as *Neofiber alleni*. The measurements of this specimen are given below in the discussion of *Neofiber* cf. *N. diluvianus*.

ChM PV5071 (not figured) is a left lower second molar of *N. alleni* found in the Ten Mile Hill Beds by Matthew Swilp in a drainage ditch beside the Baptist Mission on Trolley Road (County Road 199) in Dorchester County, ca. 21 mi. (13 km) northwest of Charleston (USGS Stallsville 7.5' quadrangle).

As summarized by Frazier (1977:372), *Neofiber alleni* seems to be a late Illinoian descendant of *N. leonardi* Hibbard, an extinct middle Pleistocene form known from Yarmouthian (Hibbard and Dalquest, 1973) deposits (Stages 8-12 of Richmond and Fullerton, 1986) in Kansas and Texas and from local faunas in Florida and West Virginia that are considered to be of late Kansan or Yarmouthian age (Frazier 1977). A third form, *N. diluvianus* (Cope), from Yarmouthian cave deposits in Pennsylvania, is poorly known and its relationships are not clear (Frazier, 1977). *Proneofiber*, a cricetid rodent from the Lower Pleistocene of Texas, has been proposed as the probable ancestor of *Neofiber* (Hibbard and Dalquest, 1973).

Frazier's (1977:371) analysis of the morphology of the lower first molar in the Irvingtonian-early Ranchlabrean *Neofiber leonardi* and the Rancholabrean-Recent *Neofiber alleni* "strongly suggests genetic continuity from *N. leonardi* through fossil *N. alleni* to recent *N. alleni*." His record of *N. leonardi* in the middle Pleistocene of Florida, coupled with numerous records of *N. alleni* in the late Pleistocene of that state (Webb, 1974a, Table 2.1), provides stratigraphic evidence indicating a *leonardi-alleni* lineage. Frazier noted that fossil specimens of the ml of *N. alleni* "show a greater tendency (approximately 50 percent) toward the primitive (anteroposterior) orientation of the third labial reentrant angle than Recent *N. alleni*." This tendency is clearly evident in the ml of ChM GPV2018, the third reentrant angle being strongly oriented anteroposteriorly (Figure 37c). In all other respects it agrees with the morphology of the ml of *N. alleni* as figured by Frazier (1977, fig. 1) and most nearly matches his figure 1-G. Most of the anterior loop is missing from the occlusal surface of the crown but the shape of the loop can be seen in its preserved basal portion.

In his discussion of the paleoecology of *Neofiber*, Frazier (1977:372) notes that "*Neofiber alleni* inhabits heavily vegetated freshwater bogs, marshes, swamps, lake edges, streambanks, and brackish river deltas" within its present range and that its apparent ancestor, *N. leonardi*, "probably lived from the Great Plains eastward as a waterweed feeder requiring permanent bodies of water able to sustain emergent vegetation. Subsequently, the range of the genus retracted to Florida and Georgia." He further states that *N. alleni* "depends on abundant aquatic vegetation" and the "greater aridity, or freezing, even seasonally, could be fatal." Hibbard et al. (1965:515) included *Neofiber* among several mammalian genera that demonstrated "southward and southwestward retractions of ranges of ecological significance during the later Pleistocene."

The Pleistocene deposits that furnished the Charleston County specimen (ChM GPV2018) were mapped by Malde (1959) as part of his Ladson Formation but are now included in the Wando Formation (Late Pleistocene) of McCartan et al. (1980). Although the original ground surface at the *Neofiber* locality was obliterated during grading for the shopping center and its parking lot, contours on the map of the Ladson quadrangle indicate an elevation of 10 to 15 feet (3.1 to 4.6 m) in the area where the specimen was found, placing these deposits well within the elevational range of the Wando Formation. Samples from a U.S. Geological Survey power auger hole on the south side of the shopping center compare favorably with the lithology of fluvial facies of the Wando Formation as described by McCartan et al. (1980). The log of this hole shows 23 feet (4.6 m) of Pleistocene sediments overlying the Ashley Formation (Late Oligocene), a marine unit underlying the entire Charleston area (personal communication, R.E. Weems, U.S. Geological Survey).

From the area in which the *Neofiber* mandible was found, the ground slopes southward to nearby Peters Creek, a small stream that flows into Goose Creek Reservoir 0.72 km east of the *Neofiber* locality. Today, this stream is only about three meters wide where it flows beneath U.S. Route 52 next to the *Neofiber* locality, but the width of the valley walls along the contours suggests that Peters Creek is the remnant of a small inlet that extended westward from Goose Creek when that tributary was a larger estuarine stream during Sangamonian time.

Having been found in fluvial deposits near a former shoreline of the Peters Creek inlet, the *N. alleni* mandible appears to document the presence of a round-tailed muskrat population along the shores of the inlet during Wando time. As noted above, coral dates from the Wando Formation (Szabo, 1985) indicate that this unit was deposited during the Sangamonian interglacial stage. A Sangamonian age for this specimen fits the historical zoogeography of *N. alleni* quite well in view of subsequent climatological developments in the Pleistocene. If the final retreat of *N. alleni* to its present range was precipitated by declining temperatures during the early stages of the Wisconsinan, *Neofiber* remains in South Carolina would be expected to predate the Wisconsinan, as would Ray's (1967) record of *N. alleni* from Pleistocene fissure fillings at Ladds, Bartow County, Georgia.

In summary, the South Carolina specimens represent the northernmost records of *Neofiber alleni* on the Atlantic Coastal Plain beyond its present range (Florida and extreme southeastern Georgia). They suggest that this species occurred in suitable habitats in the southeastern United States through the Sangamonian interglacial stage but disappeared over most of this region presumably when warm-adapted vegetative types upon which it fed were greatly reduced by changing climatic conditions associated with Wisconsinan glaciation. The two molars (SCSM, 93.105.153, 93.105.154) from the Ardis local fauna in Dorchester County do, however, provide evidence that some *Neofiber* populations were still present on the Coastal Plain of South Carolina in the Late Wisconsinan.

Neofiber CF. *N. diluvianus* (Cope, 1896)
Figures 33, 36, 37b, d-e

MATERIAL.—ChM PV5070, partial right mandibular ramus with i1, m1, m2, and root of m3 in place but missing posterior region; ChM PV5398, partial right mandibular ramus with base of i1, m1, and m2 in place but missing posterior region.

LOCALITY AND HORIZON.—North Carolina, Brunswick Co.; Calabash; excavation for Marsh Harbor Marina on Intracoastal Waterway; Robert L. Johnson; PV5070 collected in fall 1986, PV5398 in November 1994 ; Waccamaw Formation (upper bed), Early Pleistocene.

AGE.—Early Irvingtonian, Pre-Illinoian.

DISCUSSION.—In the fall of 1986 Robert L. Johnson of North Myrtle Beach, S.C., collected a small rodent jaw (Figures 36a-b, 37e) on spoil piles accumulated during excavations for the Marsh Harbor Marina in the small coastal city of Calabash at the extreme southeastern end of Brunswick County, North Carolina, a

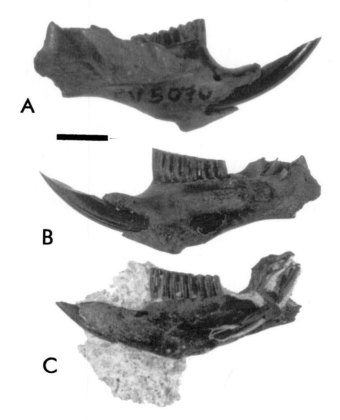

Figure 36. Partial right mandibular rami of *Neofiber* cf. *N. diluvianus* (ChM PV5070, PV5398); N.C., Brunswick Co., Waccamaw Fm. (Early Pleistocene). **A**, PV5070, labial view, i1, m1, and m2 in place; **B**, lingual view; **C**, PV5398, lingual view, m1, and m2 in place; fragment of Waccamaw Formation matrix adhering to specimen. Scale bar = 5 mm. For occlusal view see Figure 37.

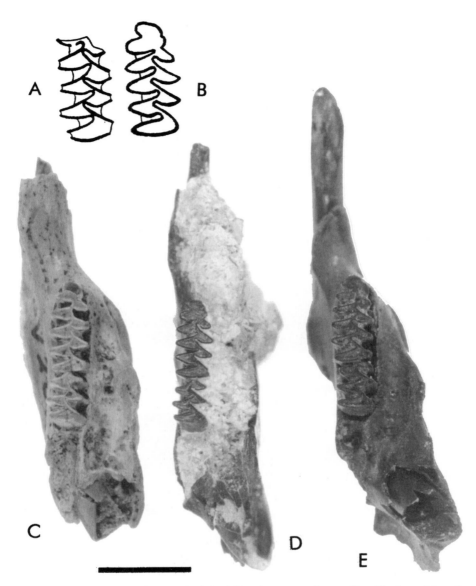

Figure 37. A, holotype right m1 of *Schistodelta sulcata* Cope (= *Neofiber diluvianus* [Cope]) (ANSP 140), Port Kennedy Cave, Pa. (Middle Pleistocene), occlusal pattern; x 6 (from Hibbard, 1955, fig. 1b). **B**, right m1, *Neofiber* cf. *N. diluvianus* (ChM PV5398), N.C., Brunswick Co., Waccamaw Fm. (Early Pleistocene); occlusal pattern; x 7. **C**, right mandibular ramus, *Neofiber alleni* (ChM GPV2018), occlusal view; S.C., Charleston Co.; Wando Fm. (Late Pleistocene). **D**, ChM PV5398, and **E**, ChM PV5070, right mandibular rami, *Neofiber* cf. *N. diluvianus*, occlusal views; N.C., Brunswick Co., Waccamaw Fm. (Early Pleistocene). Scale bar = 5 mm. (Note greater width of reentrant angles in m1 of *N. alleni* [**C**] compared to those of *N. diluvianus* [**D, E.**])

short distance northeast of the South Carolina-North Carolina state line (Figure 31). The entrance to the marina is on N.C. Route 179 near the western perimeter of Calabash 4.8 miles (7.72 km) east of U.S. Route 17, and the spoil piles are located 0.3 mile (0.48 km) south of the entrance. He brought the specimen, a partial right mandible, to The Charleston Museum for identification, and when told of its significance as the first known record of round-tailed muskrats (*Neofiber*) in North Carolina, Mr. Johnson graciously contributed the specimen to the museum collection.

Subsequently, Mr. Johnson donated another *Neofiber* mandible (ChM PV5398, Figures 36c, 37b-d) that he collected at Calabash in November 1995. Also a partial right mandible, it is firmly cemented to a small fragment of white, shelly, calcareous matrix and has been left that way to document its stratigraphic origin. In this specimen, most of the anterior portion of the incisor is missing, and, as in PV5070, the condyle and the coronoid and angular processes also are not present. The enamel pattern of m1 and m2 is almost identical to that of m1 and m2 in PV5070, varying only in the degree to which the reentrant angles are compressed.

The most nearly complete of the two specimens, ChM PV5070 is a partial right mandibular ramus with i1, m1, m2, and the base of m3 preserved. The anterior portion of the symphyseal suture is missing, and the posterior portion of the ramus—including the condyle and the coronoid and angular processes—has been broken off above and behind the posteriormost extent of the alveolus for the incisor. The specimen measures 29.5 mm from the anterior tip of the incisor to the preserved portion of the posterior region of the ramus and 11.70 mm in greatest depth (occlusal surface of m1 to ventral surface of ramus at posterior end of symphysis). Anteriorly, the axis of i1 is medial to the molars, but i1 then extends downward beneath the three molars and curves upward behind and lateral to m3, terminating in the base of the ascending process. This configuration of the incisor is a distiguishing character of the genus *Neofiber*. The reentrant angles of m1 are greatly compressed, of comparable width lingually and labially, and are anterolaterally directed (Figure 37b). The anterior loop has one enamel fold lingually and a smaller one labially. The base of m3 is present well down in the alveolus, which is missing its posterior margin. The molars have no roots.

The enamel pattern of m1 in both of these specimens most closely resembles that of *Neofiber diluvianus* (Cope, 1896), from Middle Pleistocene sediments in Port Kennedy Cave, Montgomery County, Pennsylvania. This poorly known form was originally described by Cope (1896) as *Microtus diluvianus* on the basis of the right and left M1 and M2 (ANSP 144) of a young individual. Hibbard (1955:93) referred that species to *Neofiber* and synonymized Cope's (1899:206) *Schistodelta sulcata*—based on a nearly complete right m1 (ANSP 140, Figure 37a) from the Port Kennedy Cave—with *N. diluvianus*. He observed that "The upper teeth of *Neofiber diluvianus* are distinguished from those of ... *Neofiber alleni* True by their narrower alternating triangles with more pointed apexes. The lingual reentrant angles of M^1 and M^2 in [*N. alleni*] are wider." Hibbard (1955:93) also noted that "*N. leonardi* Hibbard ...

differs from *N. diluvianus* in that the first alternating triangle of M^1 is larger and its apex is broader."

The two mandibular rami from Calabash (ChM PV5070 and PV5398) resemble each other so closely as to leave virtually no doubt that they represent the same taxon (Figures 36, 37d-e), and both are so similar to the holotype right m1 of Cope's (1899:206) *Schistodelta sulcata* (=*N. diluvianus*) (Figure 37a)—the best specimen in the type series—that it is tempting to refer them to *N. diluvianus* without hesitation. However, in the absence of a larger sample of *N. diluvianus* lower first molars for comparison and the correspondingly poor understanding of that species, a provisional assignment of the two Calabash specimens to *N. diluvianus* seems best at this time.

Measurements of the *N. alleni* mandible (PV2018), the mandibles of *N.* cf. *diluvianus* (PV5070, PV5398), and the holotype m1 of *N. diluvianus* (ANSP 140) from Port Kennedy Cave are shown in Table 6.

There are significant differences in size and in the enamel pattern of the teeth of the specimens of *N.* cf. *diluvianus* (PV5070, PV5398) and the Charleston County specimen of *N. alleni* (PV2018), which is also more robust than either of the Calabash specimens. In vertical dimension, the mandible in both forms is deepest from the dorsal margin of the ramus at m1 to the ventral surface of the posterior terminus of the symphysis. From that point the ramus narrows posteriorly toward m2. As seen in Table 6, the lingual depth of the *N. alleni* specimen (PV2018) is 8.30 mm at m1 and 7.00 mm at m2, a reduction of 16% between the two points. In the better specimen of *N.* cf. *diluvianus* (PV5070) the ramus is 7.75 mm deep at m1 and 5.50 at m2, a difference of 29%. In PV5398 the measurements are 6.55 at m1 and 5.15 at m2, a 21% difference. The two specimens of *N.* cf. *diluvianus* are smaller than the *N. alleni* mandible (ChM PV2018) and could possibly belong to relatively young animals, but the likelihood of finding two subadult mandibular rami so close in size seems rather remote, so they are regarded here as those of adult individuals. In general aspect, the two specimens of *N.* cf. *diluvianus* and the Port Kennedy specimen of *N. diluvianus* are more slender and not as heavily built as the mandible of *N. alleni*, indicating that *N. diluvianus* was a smaller form than its later Pleistocene relatives

When PV5070 was received it was free of matrix, so its actual stratigraphic origin was not apparent. Fortunately, the second specimen from the Calabash locality (ChM PV5398) was imbedded in a fragment of matrix similar in appearance to the Waccamaw Formation.. The Waccamaw is regarded by some authors to be of late Pliocene (e.g. Campbell and Campbell, 1995) or early Pleistocene (e.g. Cronin, 1990) age.

Wishing to verify the age of the matrix, I visited the Calabash site on 3 September 1995 and made a small collection of matrix samples and mollusks from the many specimens scattered about on the spoil piles, which consisted entirely of material dug from the Waccamaw Formation. Specimens of the following molluscan taxa were collected:

Bivalves	N=	ChM No.
Anadara lienosa (Say)	5	PI16080-PI16084
Glycemeris americana (DeFrance)	4	PI16085-PI16088
Crassostrea virginica (Gmelin)	5	PI16100-PI16104
Pleuromeris tridentata (Say)	1	PI16093
Mercenaria mercenaria (Linnaeus)	3	PI16076-PI16078
Chione grus (Holmes)	4	PI16089-PI16092
Gastropods		
Busycon contrarium (Conrad)	1	PI16099
Crepidula fornicata (Linnaeus)	3	PI16094-PI16096
Aurinia obtusa (Emmons)	1	PI16098
Conus waccamawensis (Smith)	1	PI16097

Because the bivalve *Anadara lienosa* and the gastropod *Conus waccamawensis* are not known from beds younger than the Waccamaw (Lyle D. Campbell, personal communication, 1996), the spoil pile material clearly was dug from the Waccamaw Formation, and the presence of the *Neofiber* mandible (PV5398) firmly imbedded in Waccamaw matrix leaves virtually no doubt that this specimen is of Waccamaw age. Campbell and Campbell (1995:66–67) have noted that "the upper Waccamaw is separated from the lower Waccamaw by an unconformity at Calabash." Therefore, the stratigraphic origin of the spoil pile material is of utmost importance, because Campbell and Campbell (1995:67) observed that "the lower Waccamaw at Calabash contains planktic foraminifera indicative of an age of 2.4 Ma (Huddleston, 1975, personal communication)." Relating that date to the Krantz (1991) model of Pliocene sea level fluctuations, Campbell and Campbell (1995:67) concluded that "The consequent age for the upper Waccamaw would be 1.9 to 2.2 Ma." They further observed that "The upper Waccamaw directly correlates with the upper Caloosahatchee of southern Florida, which directly underlies the 1.5 Ma Bermont beds. . . . The Bermont fauna in Florida is essentially modern (85 percent extant). . . . The James City-upper Waccamaw-upper Cypresshead-lower Nashua-upper-Caloosahatchee faunas form a zoogeographic continuum of warm temperate to subtropical to tropical molluscan species only approximately 50 percent extant" (Campbell and Campbell, 1995:67). The absence of the bivalves *Carolinapecten eboreus* and *Amusium mortoni* among the mollusks found on the spoil piles from the marina at Calabash is significant, because these two forms occur in the lower bed but not in the upper bed, as indicated by an earlier collection of mollusks from both beds at Calabash made by Evelyn Dabbs during the summer of 1985 and subsequently placed in The Charleston Museum (ChM PI11485-11698). When they are present in a molluscan assemblage, both *C. eboreus* and *A. mortoni* are usually fairly common but were not evident at all when we collected the recent sample of shells from the spoil piles that produced the *Neofiber* mandible. Thus, the spoil pile material seems clearly to have been obtained from the upper Waccamaw bed, which, according to Campbell and Campbell (1995:67), would place the age of the *Neofiber* mandible at 1.9 to 2.2 Ma.

Such an early date is not, however, consistent with the temporal distribution of microtine (= arvicoline) rodents with rootless teeth, e.g., *Neofiber*. As Repenning (1987:249) has observed, "Within the endemic fauna of the United States, most older forms with rooted teeth become extinct by Irvingtonian I time [1.9-ca. 0.9 Ma], and microtine faunas of this age are characterized by forms with rootless teeth." Approximately 90% of microtine (=arvicoline) rodents with rooted teeth had become extinct before the end of the Pliocene (C.A. Repenning, personal communication, July 2002); hence, the presence of *Neofiber*—a form with rootless teeth—in the upper Waccamaw bed lends strong support to proposals that this unit is of early Pleistocene age (e.g., Gibson, 1983:38, fig. 2; Owens,1989; Cronin, 1990), and that concept is accepted in this paper.

As noted in the *N. alleni* account above, *Proneofiber*, from early Pleistocene deposits at Gilliland, Texas, was proposed as the probable ancestor of *Neofiber* by Hibbard and Dalquest (1973). In their respective chronological charts, Lundelius et al. (1987) and Repenning (1987) placed the Gilliland fauna at an age level of approximately 1.2 Ma. With better early specimens of *Neofiber* now available from the upper bed of the Waccamaw Formation, dated at 1.37-1.2 Ma (Bybell, 1990), the relationships between *Proneofiber* and those specimens, reported herein as *N.* cf. *N. diluvianus*, are in need of investigation, inasmuch as the latter specimens appear to be as old or slightly older than the proposed ancestor of *Neofiber*. They also appear to be older than the Port Kennedy fauna. Kurtén and Anderson (1980:33) noted the presence of *Smilodon gracilis* in the Port Kennedy fauna, and *S. gracilis* also occurs in the Hamilton Cave fauna, suggesting a similar age for the two faunas, the latter having been dated at about 840,000 ypb on the basis of the microtine (= arvicoline) rodent fauna (Reppening, 1992; Reppening et al., 1995). If the Port Kennedy fauna is of similar age to that of Hamilton Cave, the Calabash specimens are some 450,000 years older than the type material of *N.diluvianus*, but that expanse of time does not mitigate against referral of the Calabash material to *N. diluvianus*. As listed by Repenning (1987:260-261; fig. 81), the microtine (= arvicoline) rodent *Synaptomys kansanensis* first appears in the El Casco, California, fauna (c. 1.85 Ma.) and is last seen in the Alamosa, Colorado, fauna (0.84-.70 kya), approximately one million years later.

The presence of *Neofiber* in North Carolina well before the beginning of Middle Pleistocene glaciation and the accompanying climatic changes is compatible with the general distribution patterns of the genus. *Neofiber* is absent in the Leisey Shell Pit Local Fauna from the early Pleistocene Bermont Formation of Florida (Morgan and White, 1995:447, table 8), in which the presence of the muskrat *Ondatra annectens* would seem to indicate suitable *Neofiber* habitat, and its absence in other early Irvingtonian faunas from Florida suggests that *Neofiber* may not have reached Florida until middle Irvingtonian time, small mammal remains of that age from the McLeod Limerock Mine in Levy County having furnished the first and only record of *Neofiber leonardi* from that state (Frazier, 1977:328). Thus, the occurence of *Neofiber diluvianus* in southern Pennsylvania and *Neofiber* cf. *N. diluvianus* in upper Waccamaw sediments on the southernmost coast of North Carolina may indicate that the range of this genus along the Atlantic Coast did not extend south of the Carolinas in Early Pleistocene time.

* * *

Suborder HYSTRICOMORPHA
Family ERETHIZONTIDAE Thomas, 1896
Subfamily ERETHIZONTINAE Thomas, 1896
Genus ERETHIZON Cuvier, 1822
ERETHIZON DORSATUM (LINNAEUS, 1758)
Figures 31, 38

MATERIAL.— DM 1, left mandibular ramus missing the anterior end and the posterior region behind m2 and with p4, m1, m2, and the root of the incisor in place; SCSM 83.17.1, left p4.

LOCALITY AND HORIZON.—DM 1: S.C., Charleston Co.; Edisto beach; Gracie Marvin, 11 March 1991. SCSM 83.17.1: S.C., Charleston Co.; Edisto beach; Margaret Pulliam, c. 1983. Undetermined offshore unit. Upper Pleistocene.

AGE.—Rancholabrean, Late Pleistocene.

DISCUSSION.—Although not among the Recent mammal fauna of the lower southeastern United States, porcupines have been recorded from several Late Pleistocene localities in Florida (Frazier, 1981) and thus would not be unexpected in the Pleistocene of South Carolina, but the present specimens are the first known records of *Erethizon* in this state. They also appear to be the only records of porcupines on the Atlantic Coastal Plain between Florida and Virginia.

In DM 1 (Figure 38) the anterior portion of the symphysis and the exposed (anterior) length of the incisor are broken off. The posterior region of the ramus is also gone, along with m3, but p4, m1, and m2 are present and are well preserved. The isolated left p4 (SCSM SC83.17.1; not figured) is well preserved and shows only moderate occlusal wear.

Measurements (in mm) are as follows:

	DM 1	SCSM 83.17.1
Anteroposterior length of ramus as preserved	40.7	——
Depth of ramus at m2	18.0	——
Transverse diameter of ramus below p4	7.8	——
Greatest transverse diameter of ramus	12.6	——
Width of p4	6.4	4.55
Width of m1	6.1	——

As shown above, the isolated p4 (SCSM 83.17.1) from Edisto Beach is considerably smaller than its counterpart in the mandible (DM 1) from the same locality and more nearly conforms to the range of variation in the width of p4 in the more diminutive porcupine genus *Coendou* (Frazier, 1981: fig. B-16) and in *Erethizon kleini*, described by Frazier (1981:43) from the Early Irvingtonian Inglis IA site in Citrus County, Florida. *E. kleini* is distinguished from *E. dorsatum* by its smaller size and is the earliest record of the genus in eastern North America, probably becoming

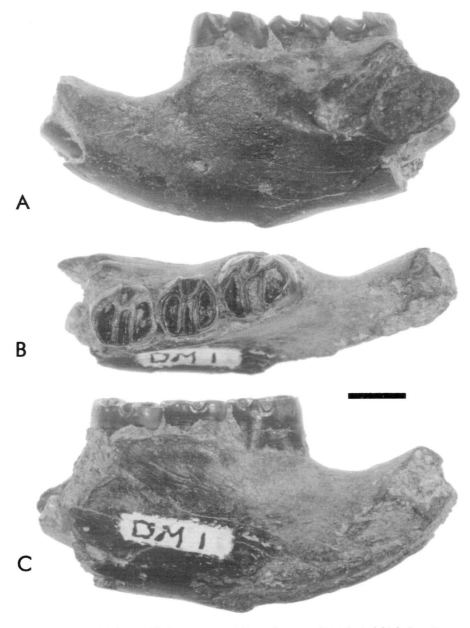

Figure 38. Partial left mandibular ramus, *Erithizon dorsatum* (DM1). **A**, labial view; **B**, occlusal view; **C**, lingual view. S.C., Charleston Co.; undetermined offshore unit (Late Pleistocene). Scale bar = 5 mm.

extinct during Irvingtonian time (Frazier,1981:50). If *E. kleini* did not survive into the Rancholabrean it thus seems unlikely that SCSM 83.17.1 is referable to that species because beds deposited during the Irvingtonian Land Mammal Age do not occur in the immediate vicinity of Edisto Island. The small size of this tooth would simply indicate that it came from the mouth of a young individual of *E. dorsatum*.

According to Frazier (1981:44), the modern species *E. dorsatum* appeared in the middle Irvingtonian and is the form to which all subsequent North American fossil porcupine remains are referable. In the eastern United States, porcupines do not presently occur south of Virginia and northeastern Tennessee (Hall and Kelson, 1959:782).

* * *

Family HYDROCHOERIDAE
Subfamily HYDROCHOERINAE
Genus *Hydrochoerus* Brisson, 1762
HYDROCHOERUS HOLMESI SIMPSON, 1928
Figures 39, 40

MATERIAL.—ChM GPV1303, left M3 with fragments of maxilla adhering; ChM GPV624, left M1; ChM PV5839, partial right m3; ChM PV5930, anterior portion of right M3; ChM PV4799, anterior portion of left M3; ChM PV2507, anterior portion of right M3; ChM PV4506, partial left p4; ChM PV4795, anterior two-thirds of left M3; ChM PV4796, posterior portion of left M3.

LOCALITY AND HORIZON.—ChM GPV1303: S.C., Horry Co.; fill material in driveway, Ocean Lakes Campground, U.S. Rt. 17 c. 8.0 km south of Myrtle Beach; Doris Holt, 15 August 1973; Waccamaw Formation, Early Pleistocene. ChM GPV624: S.C., Charleston Co.; North Charleston; spoil pile behind shopping center on Rivers Avenue (U.S. Route 52) opposite Aviation Avenue; Doris Holt, August 1974; Ten Mile Hill Beds, Late Middle Pleistocene. ChM PV5839: S.C., Dorchester Co.; ditch in Irongate subdivision, Vance McColllum, summer, 1980; Ten Mile Hill Beds, Late Middle Pleistocene. ChM PV5930: S.C., Charleston Co.; North Charleston, Hawthorne Trailer Park, east side Rivers Avenue near Aviation Avenue; Doris Holt, summer 1975; Ten Mile Hill Beds, Late Middle Pleistocene. ChM PV4799: S.C., "South Carolina phosphate beds;" Earl Sloan, ca. 1900; stratigraphic origin uncertain; Late Pleistocene. ChM PV2507: S.C., Charleston Co.; vicinity of Runnymede Plantation; C.C. Pinckney, Jr., ca. 1900; Wando Formation, Late Pleistocene. ChM PV4506: S.C., Charleston Co.; borrow pit for Mark Clark Expressway, E. side of S.C. Route 61 near Ashley Hall Plantation Road; Doris Holt, December 1979; Wando Formation, Late Pleistocene. ChM PV4795, PV4796: Georgia, Chatham Co.; near Savannah; Ivan R. Tomkins, 22 March 1936; formation undetermined, Late Pleistocene.

AGE.—ChM GPV1303: Early Irvingtonian, Pre-Illinoian. GPV624, PV5839, PV5930: Early Rancholabrean, Late Illinoian. PV4799: Rancholabrean, Illinoian-Sangamonian. PV2507, PV4506: Middle Rancholabrean, Sangamonian.

DISCUSSION.—Capybaras were first recorded from South Carolina by Leidy (1853:241), who reported a fragment of an incisor "of a large Rodent animal allied to *Hydrochoerus capybara*" collected by Francis S. Holmes along the Ashley River near Charleston. To this specimen he applied the name *Oromys Æsopi*. Holmes most likely found the tooth at his principal spot for collecting Pleistocene fossils on the Ashley River, the west bank of the river at the old Bee's Ferry landing, discussed above as the type locality for *Arctodus pristinus*. Subsequently, Leidy (1856:165) re-

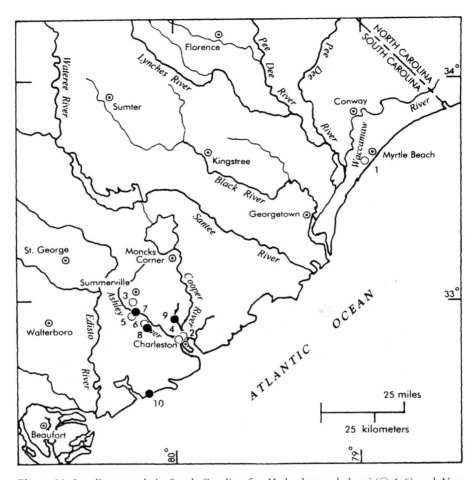

Figure 39. Locality records in South Carolina for *Hydrochoerus holmesi* (○ 1-6) and *Neochoerus pinckneyi* (● 7-10). **1**, ChM GPV1303, Horry Co., Ocean Lakes Campground; Waccamaw Fm. (Early Pleistocene); **2**, ChM GPV624, Charleston Co., North Charleston, spoil pile on Rivers Avenue (U.S. Route 52), Ten Mile Hill Beds (Late Middle Pleistocene); **3**, ChM PV5839, Dorchester Co., ditch in Irongate subdivision, Ten Mile Hill Beds (Late Middle Pleistocene); **4**, ChM PV5930, Charleston Co., North Charleston, Hawthorne Trailer Park, Ten Mile Hill Beds (Late Middle Pleistocene); **5**, ChM PV2507, Charleston Co.; vicinity of Runnymede Plantation, Wando Fm. (Late Pleistocene); **6**, ChM PV4506: S.C., Charleston Co.; borrow pit for Mark Clark Expressway, S.C. Route 61, Wando Fm. (Late Pleistocene). **7**, Type locality; ChM PV2506 (holotype), PV2508, PV2509, PV2510, Charleston Co., dredged from Ashley River in vicinity of Runnymede Plantation, S.C. Route 61, Goose Creek Limestone (Mid Pliocene); **8**, ChM PV4505, Charleston Co., borrow pit for Mark Clark Expressway, S.C. Route 61, Goose Creek Limestone (Mid Pliocene); **9**, ChM 2796, Berkeley Co., bottom of West Branch of Cooper River Wando Fm. (Late Pleistocene); **10**, ChM PV2439, PV2440, Charleston Co., Edisto Beach, undetermined offshore unit (Late Pleistocene).

ferred his *O. Æsopi* to the genus *Hydrochoerus* on the basis of two molar teeth from the Ashley River in the collection of Captain A.H. Bowman of Charleston, noting that these teeth came from "the same locality" as the incisor. In a later account, Leidy (1860:112–113; pl. 21, figs. 3–6) reviewed the three original specimens, reported a

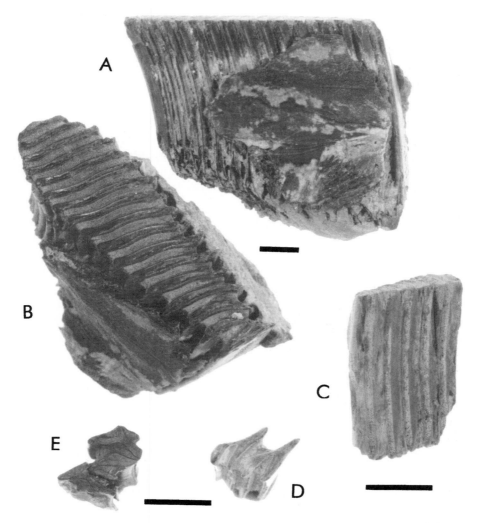

Figure 40. Teeth of *Hydrochoerus holmesi* from South Carolina. **A**, ChM GPV1303, left M3 with fragments of maxillla adhering; labial view; **B**, occlusal view; Horry Co., Waccamaw Fm. (Early Pleistocene). **C**, ChM PV5930, posterior end of right M3, lingual view; Charleston Co., Ten Mile Hill Beds (Late Middle Pleistocene); **D**, ChM GPV624, fragment of left M1, occlusal view; Charleston Co., Ten Mile Hill Beds (Late Middle Pleistocene). **E**, ChM PV4506, fragment of left P4, occlusal view; Charleston Co., Wando Fm. (Late Pleistocene). Scale bars = 10 mm.

third molar fragment obtained by Holmes "from the Ashley [River] Post-Pleiocene deposit," figured them (figs. 4–5), and indicated the relative positions of the molar fragments by shading the appropriate elements in a figure of the left mandible of modern *Hydrochoerus hydrochaeris* (fig. 6). By the time that Francis Holmes sold his collection of fossils to the American Museum of Natural History in 1872, the two Bowman specimens had become a part of the Holmes collection, uniting the holotype incisor (AMNH 485) with the other three specimens reported and figured by Leidy (1860) as *Hydrochoerus aesopi*. The two molars in the Bowman collection—

the last two prisms of a left m1 and the first prism of a left m2—were thought to have come from the same individual and were catalogued as AMNH 101240. The third molar fragment, collected by Holmes on the Ashley River, was described by Leidy (1860:113) as "the anterior pair of columns of the third tooth of the left series of the lower jaw," inferring it to be the anterior portion of an m2, the third tooth in the molariform series in hydrochoerids. However, Leidy's (1860) figures 5 and 6 (plate 21) depict this fragment as the first prism of a left m3, matching AMNH 101239, the fourth specimen in the published syntypic series. Evidently, in calling it "the third tooth of the left series," Leidy (1860:113) inadvertently used the word "tooth" instead of "molar." A fifth specimen in the *H. aesopi* material, a chip from an incisor (AMNH 101241), seems to have been overlooked in the literature.

Leidy (1860:112) observed that the enamel surface of the holotype incisor (AMNH 485) "is more strongly ridged, longitudinally, than in any of the specimens with which it was compared." In her unpublished thesis reviewing the Hydrochoeridae, Ahearn (1981:55) also called attention to that feature, noting that it is unlike any known species of *Hydrochoerus* but is similar to the ridging on the incisors of many specimens of *Neochoerus*. Ahearn (1981:55) has suggested that *H. aesopi* should be considered as a *nomen dubium*, and Spamer et al. (1995:163) have proposed that *Oromys Æsopi* is a *nomen nudum*. Considering the similarities in morphology between the holotype incisor of *H. aesopi* and the incisor of *Neochoerus*, indicating its more appropriate allocation to *Neochoerus*, it would appear that the name *Hydrochoerus aesopi* should at least be regarded as a junior synonym of *Neochoerus pinckneyi*, the species commonly occurring in the region in which the *H. aesopi* type was found. Because the three molar teeth of the syntypic series were not found in association with the holotype incisor (AMNH 485), their proper allocation is questionable in any case, and these specimens should be compared with appropriate elements of the dentition of *Hydrochoerus holmesi* and *Neochoerus pinckneyi*.

During the two weeks that O.P. Hay spent at The Charleston Museum in 1915 examining Pleistocene mammal remains in our holdings and some private collections around the city, a small collection of vertebrate fossils from the phosphate beds was donated to the Museum by former state geologist Earl Sloan of Charleston. Hay himself made out the handwritten catalogue cards in use at the time and identified three capybara teeth as *Hydrochoerus Æsopi*. Those specimens, all portions of upper third molars, were initially catalogued as ChM 13485a, b, and c but have been recatalogued as PV4798 (13485b), PV4799 (13485a), and PV4800 (13485c). Later, Hay (1923b:104) referred those specimens to his new taxon *Hydrochoerus pinckneyi* (Hay, 1923a:365). Reexamination of those teeth during the present study reveals that one of them, PV4799, is referable to *H. holmesi*. *Hydrochoerus holmesi* has been recorded from numerous Late Blancan to Late Rancholabrean sites in Florida (Kurten and Anderson, 1980:274; Ahearn, 1981:59–60; Morgan and White, 1995:429). Ahearn (1981:59-60) listed a left M3 (USNM 11822), a left M3 fragment (USNM 187210), and a right m3 (USNM 187209) from the Savannah River, Chatham County, Georgia, and two specimens from South Carolina, *viz.*, a right M3 (AMNH 32599) from the Ashley River in Charleston County and a left M3 (USNM

181573) from Darlington in Darlington County. Those unpublished records were the first indications of the presence of *H. holmesi* north of Florida. Several specimens that have accumulated in the Charleston Museum collection provide additional evidence of this species in Georgia and South Carolina. Separation of teeth of *H. holmesi* from those of *Neochoerus pinckneyi* has been accomplished through comparison of the transverse diameters of teeth in a skull and mandibles of the Recent *Hydrochoerus hydrochaeris* (ChM CM1545) with those of a complete set of the lower right molariform dentition of *N. pinckneyi* (ChM PV2796) (Table 7).

In August 1975, a well-preserved left M3 (ChM GPV1303, Figures 40a-b) of *H. holmesi* was found by Doris Holt in fill material in a driveway at the Ocean Lakes Campground near Myrtle Beach, Horry County, South Carolina. Fragments of the maxilla are still attached to this tooth. Matrix adhering between the lamellae laterally and ventrally contains very small subrounded quartz grains virtually identical to those seen in the matrix cemented to the mandible of *Neofiber* cf. *N. diluvianus* (ChM PV5398) from the Waccamaw Formation. Richard E. Petit (personal communication, April 1998), a long-term resident of North Myrtle Beach, confirms that for many years truckloads of Waccamaw material have been used as roadfill for driveways in the coastal region about Myrtle Beach. Thus, there seems to be no doubt that the specimen came from the Waccamaw Formation, placing *H. holmesi* in South Carolina in early Irvingtonian time.

TABLE 7. Measurements (in mm) of transverse diameter of upper and lower right molarifor teeth of modern *Hydrochoerus hydrochaeris* (ChM CM1545), Pleistocene specimens of *Hydrochoerus holmesi* from South Carolina, upper and lower third molars of *Neochoerus pinckneyi*, and right M3 and lower right dentition of a single individual of *N. pinckneyi* (ChM PV2796) from Charleston County, South Carolina.

	P4	M1	M2	M3	p4	m1	m2	m3
H. hydrochaeris, ChM CM1545)	9.0	9.4	10.3	13.0	9.8	10.3	12.0	14.4
H. holmesi, ChM PV624 left	10.0							
" ChM GPV1303, left				16.7				
" ChM PV4795, left			14.8					
" ChM PV4796, left			13.7					
" ChM PV5839, right		14.0						
" ChM PV4799, right			14.9					
" ChM PV2507, right			13.0+					
" ChM PV5930, right			13.2					
" ChM PV4506, left...				9.0+				
N. pinckneyi, ChM PV2506, holotype, left				21.7				
" ChM PV2508, left				21.5				
" ChM PV2510, left				20.1				
" ChM PV4798), left				18.1				
" ChM PV2509, right				19.0				
" ChM PV2439, right				18.7				
" ChM PV4800, right				18.2				
" ChM PV4797, left								16.0
" ChM PV2796, right				16.9	10.0	10.5	13.5	15.6

Three specimens from the Ten Mile hill Beds—ChM GPV624 a left M1 (Figure 40d), and ChM PV5930, the anterior portion of right M3 (Figure 40c), both collected by Doris Holt in Charleston County, and ChM PV5839, a right m2 collected by Vance MacCollum in Dorchester County—provide evidence of *H. holmesi* in the Late Middle Pleistocene of South Carolina. A fourth specimen, ChM PV4799, a partial right M3 with no other data than the "S.C. phosphate beds," was one of the three specimens that Hay initially determined as *Hydrochoerus aesopi* and later referred to *H. pinckneyi* (Hay, 1923:103-104). It is similar in color to the yellowish white of specimens from the Ten Mile Hill Beds but has a much grayer cast and may be from some other unit, but there is no immediate way of knowing if it is from the Charleston area. The donor, Earl Sloan, had obtained vertebrate material from dredgings in the Coosaw River near Beaufort as well as specimens from the vicinity of Charleston. The Wando Formation has furnished ChM PV2507, a right M3, and PV4506, a partial left p4 (Figure 40e), both from the vicinity of the Ashley River in Charleston County. Though discolored through mineralization, the latter specimen is not black and worn like PV2507 and many of the *Neochoerus pinckneyi* specimens from the Wando deposits along the Ashley River, indicating that it was in place in the Wando beds and thus is of Sangamonian (Middle Rancholabrean) age. Three specimens, a left M3 (PV4795), a right M3 (PV4796), and a left m3 (PV4797) collected by Ivan R. Tomkins near Savannah, Chatham County, Georgia, in 1936, probably are of Sangamonian age as well.

In summary, *Hydrochoerus holmesi* is represented in the fossil record of South Carolina by AMNH 32599, a right M3, USNM 181573, a left M3, and seven teeth in The Charleston Museum that record the presence of that taxon in South Carolina from the Early Irvingtonian to the Middle Rancholabrean.

* * *

Genus *Neochoerus* Hay, 1926
NEOCHOERUS PINCKNEYI (HAY, 1923)
Figures 39, 41, 42

MATERIAL.—ChM PV2506, left M3, holotype, *Hydrochoerus pinckneyi* (Hay, 1923a); ChM PV2508, PV2509, left M3; ChM PV2510, fragment of left M3; ChM PV4505, fragment of left M1; ChM PV4798, fragment of left M3; ChM PV4800, fragment of right M3; ChM PV2796, skullcap, left and right premaxillae, right p4, m1, m2, and m3; ChM PV2439, right M3 (Roth and Laerm, 1980:14); ChM PV2440, right I1 (Roth and Laerm, 1980:14).

LOCALITY AND HORIZON.—ChM PV2506 (holotype), PV2508, PV2509, PV2510: S.C., Charleston Co.; dredged from Ashley River in vicinity of Runnymede Plantation, ca. 17.7 km (11.0 mi.) northwest of Charleston; C.C. Pinckney, ca. 1900; Goose Creek Limestone, Mid Pliocene. ChM PV4505: S.C., Charleston Co.; borrow pit for Mark Clark Expressway, S.C. Route 61 near Ashley Hall Plantation Road; Doris Holt, December 1979; Goose Creek Limestone, Mid Pliocene. ChM PV4798, PV4800: S.C., Charleston Co.; "South Carolina phosphate beds;" ?Earl Sloan, ca.

1900; Wando Formation, Late Pleistocene. ChM 2796: S.C., Berkeley Co.; bottom of West Branch of Cooper River, ca. 1.8 km (1.10 mi) south-southwest of Strawberry Chapel; Barry Albright, summer 1977; Wando Formation, Late Pleistocene. ChM PV2439: S.C., Charleston Co.; Edisto Beach; Doris Holt, summer 1978; ChM PV2440, Edisto Beach, Doris Holt, May 1978; undetermined offshore unit; Late Pleistocene.

AGE.—ChM PV2506 (holotype), PV2508, PV2509, PV2510, PV4505: Mid Blancan. ChM PV4798, PV4800, PV2796: Rancholabrean, Sangamonian. ChM PV2439, PV2440: Late Rancholabrean, Wisconsinan.

DISCUSSION.—*Neochoerus pinckneyi* has been recorded from the late Pleistocene of South Carolina by Roth and Laerm (1980:14), who reported a right M3 (ChM PV2439) and a right I1 (ChM PV2440, Figure 41c) collected on Edisto Beach by Doris Holt in 1978. This large capybara was described as *Hydrochoerus pinckneyi* by Hay (1923a:365) on the basis of a left M3 (ChM PV2506, Figures 41a-b) then in the collection of Charles C. Pinckney, Jr., owner of the Magnolia Phosphate Mine at Runnymede Plantation on the west bank of the Ashley River approximately 11.0 miles (17.7 km) northwest of Charleston.

Initially, Hay (1923a:365) stated that "exactly where the [holotype] tooth was found is not known, but it was somewhere in the vicinity of Charleston." Shortly thereafter, he published additional notes on *H. pinckneyi* in which he reported that "it was found in dredging for phosphate rock, somewhere north of Charleston, South Carolina, in probably the Ashley River" (Hay, 1923b:103). The Ashley River is, in fact, the only river north of Charleston in which dredging for phosphate was conducted; thus it seems reasonable to conclude that it was recovered during dredging of the river in the vicinity of Pinckney's mine.

Hay (1923a:365) gave the measurements of the holotype m3 as 62 mm in length, 17.5 mm in width, and "the height of the plates on the inner surface" as 37 mm. In a later paper, Hay (1923b:103) again provided measurements of the holotype, which he gave as 62 mm anteroposteriorly along the occlusal surface, "the width in front" as "very close to 63 mm; toward the rear, where widest, 22 mm," and 27 mm in height. His measurement of 63 mm in anterior width is evidently a mistake, but there is a 10-mm difference in the two measurements of the height. To clarify the discrepancies in the two sets of data, the following measurements (in mm) were taken:

Anteroposterior length along oclussal surface	62.0
Anterior width	17.6
Greatest posterior width	22.0
Vertical diameter at center, as preserved	31.55

Hay (1923b: plate 7) also included labial and occlusal views of the holotype and two similar views of a second, smaller specimen (ChM PV4798). The latter specimen is one of three fragmentary upper third molars from the "South Carolina phosphate beds" given to The Charleston Museum by Earl Sloan of Charleston, as mentioned above in the *Hydrochoerus holmesi* account. Hay (1923b:104) called attention to these specimens (ChM PV4798, PV4799, and PV4800), observing that they "appear

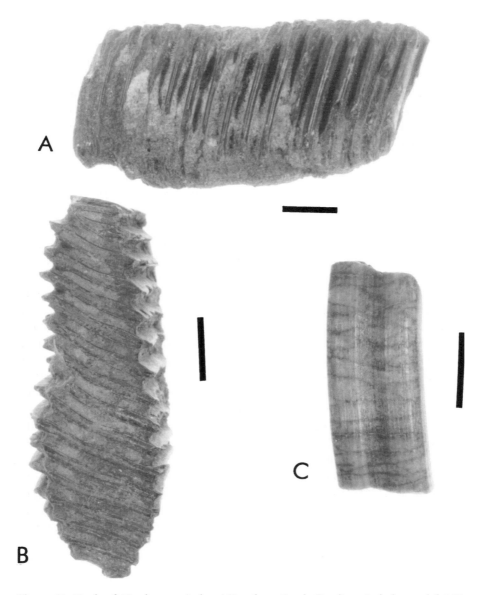

Figure 41. Teeth of *Neochoerus pinckneyi* Hay from South Carolina. **A**, holotype left M3 (ChM PV2506), labial view; **B**, occlusal view; Charleston Co., Ashley River, vicinity of Runnymede Plantation, Goose Creek Limestone (Mid Pliocene); **C**, right I1 (ChM PV2440), anterior view, showing longitudinal ridges characteristic of *Neochoerus*; Charleston Co., Edisto Beach (Late Pleistocene). Scale bars = 10 mm.

to belong to *H. pinckneyi.*" PV4798 and PV4800 are well within the range of measurements of the width of upper third molars of *N. pinckneyi* in the Charleston Museum collections (Table 7) and are within the parameters diagrammed by Ahearn (1981:fig. 6); but, as noted above, PV4799 is referable to *Hydrochoerus holmesi*, which was not described until 1928 and thus was unknown to Hay in 1923.

In a subsequent publication on some Pleistocene vertebrate remains from Texas, Hay (1926:4–7) reported a large capybara mandible that was accompanied by a piece of the left maxilla with M3 in place. The measurements and number of plates of the tooth were close enough to those of the holotype of *H. pinckneyi* to assure Hay that "the South Carolina and the Texas specimens belong to the same species," but the length and form of the masseteric ridge of the mandible differed sufficiently from the ridge in *Hydrochoerus hydrochaeris* to convince him that two distinct genera were involved. Therefore, he erected the genus *Neochoerus* and referred his *H. pinckneyi* to that taxon (Hay, 1926:7).

When the Pinckney collection was given to the Charleston Museum in 1964, three additional upper third molars of *N. pinckneyi* (ChM PV2508, PV2509, and PV2510) from the vicinity of Runnymede Plantation were among the specimens received. In all three specimens the edges of the laminae are badly chipped and broken, somewhat more so than the type specimen, suggesting that they, too, were recovered during dredging operations. In taking the measurements of these specimens I noticed that the spaces between the laminae of one of them (PV2510) were filled with a hard whitish matrix firmly cemented to the specimen. Microscopic examination revealed this matrix to contain tiny black grains of phosphate that in the Charleston area are characteristic of the Goose Creek Limestone, a Mid-Pliocene marine unit laid down approximately 3.6 million years ago (Weems et al., 1982; Campbell and Campbell, 1995, fig. 1). Examination of the other two teeth, PV2508 and PV2509, showed traces of Goose Creek matrix as well. The holotype left M3 (Chm PV2506) was then examined and found to have substantial amounts of this medium-to coarse-grained calcarenite adhering to it between the laminae and on the ventral surface. Though much of the matrix is discolored to a grayish brown, probably by tannic acid staining while the specimen was in the Ashley River, the lithic inclusions (i.e., tiny subrounded grains of quartz and black phosphate) are characteristic of the Goose Creek Limestone and can be seen also in matrix cemented to ChM PV4505, the anterior portion of a left M1 of a capybara. That specimen was collected in December 1979 by Doris Holt from the Goose Creek Limestone in the now-defunct Mark Clark Expressway borrow pit on S.C. Route 61 approximately 8.8 km (5.5 miles) south of Runnymede Plantation. There, a thin mollusk-rich layer of Goose Creek sediments unconformably overlies the Ashley Formation (Late Oligocene) and is unconformably overlain by the Wando Formation (Late Pleistocene). Campbell and Campbell (1995:64) note that the upper bed of the Goose Creek Limestone is present at this locality and chart the age of that unit as between approximately 3.75 and 3.55 million years (Campbell and Campbell (1995:55, fig. 1). With the discovery that the holotype and three other teeth of *N. pinckneyi* also came from the Goose Creek Limestone, it seems reasonable to refer PV4505 to that taxon in the absence of knowledge of another hydrochoerid species in this area in Mid-Pliocene time. As noted above, the Mark Clark pit also furnished a left p4 of *Hydrochoerus holmesi* from the Wando Formation. Although incomplete, the younger specimen is not as heavily mineralized as PV4505, is in much better condition, and has no matrix adhering to it. Characteristically, only two formations in the Charleston area produce fossils that often have matrix cemented to them—the Upper Oligocene Ashley Formation and the Mid-Pliocene Goose Creek Limestone. Both are calcarenites and are frequently highly indurated, but the presence of small black grains of phosphate easily distinguishes Goose Creek matrix from Ashley matrix. As in the case of the *N. pinckneyi* specimens, the holotype of *Ceterhinops longifrons* Leidy, 1877 (ANSP 11420), listed by Spamer et al. (1995:278) as being of Pleistocene age, also bears matrix of the Goose Creek Limestone and thus is at least Mid Pliocene in age but may have been reworked from Miocene units eroded by Pliocene seas.

During the summer of 1977 Barry Albright recovered the only cranial elements of *N. pinckneyi* from South Carolina (ChM PV2796) on the bottom of the West Branch of the Cooper River approximately 1.8 km south-southwest of Strawberry Chapel in Berkeley County (USGS Kittredge 7.5-minute quadrangle). They consist of the greater portion of the left premaxilla with the anteriormost end of the left maxilla attached, the entire tooth-bearing portion of the anterior end of the right premaxilla, and the skullcap (Figures 42a-c). The right M3 and the entire lower right molariform dentition (p4-m3) (Figures 42d-e) also were found but without any trace of the right mandible itself. Measurements (in mm) of the Cooper River cranial elements and comparable points on UF 22599, a complete skull of *N. pinckneyi* from Palm Beach County, Florida, are as follows:

	ChM PV2796	UF 22599
Anteroposterior length of left premaxilla, as preserved	124.2	not taken
Anteroposterior length of right premaxilla, as preserved	76.0	not taken
Depth of left premaxilla immediately anterior to nasal opening	42.9	43.3
Anteroposterior length of skullcap along midline (posterior margin of supraoccipital to anterior margin of parietal)	79.6	86.2
Anteroposterior length of supraoccipital along midline	25.2	31.0
Anteroposterior length of parietal along midline	57.4	55.2
Transverse diameter of parietal at anterior margin	86.9	not taken
Transverse diameter of supraoccipital at posterior margin	65.6	not taken

Although only a few comparative measurements were possible between the two specimens, they do not differ dramatically and thus are sufficient to demonstrate that PV2796 is referable to *N. pinckneyi*. Because the specimens were found on the bottom of the river out of their original stratigraphic context, their age is problematical. However, the preservation of the PV2796 material is identical to that of an essentially complete skull of *Tapirus veroensis* (ChM PV4257) found at the same locality as PV2796.

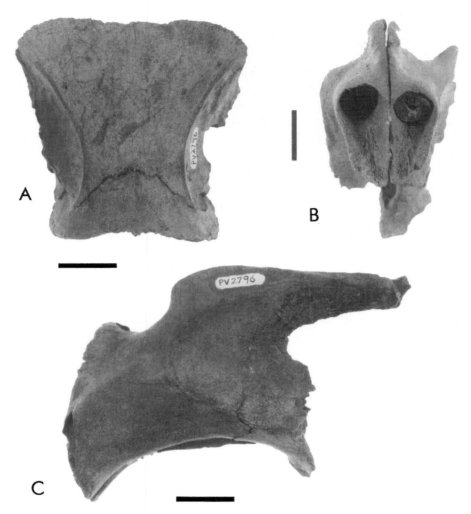

Figure 42. Partial cranium, premaxillary symphysis, and teeth of *Neochoerus pinckneyi* (ChM PV2796). S.C., Berkeley Co., bottom of West Branch of Cooper River; Wando Fm. (Late Pleistocene). **A,** partial cranium, dorsal view. **B,** premaxillary symphysis with root of left I1 in place; anterior view. **C,** left premaxilla with anterior end of maxilla, left lateral view. Scale bars = 20 mm.

In their discussion of the *Tapirus* skull, Ray and Sanders (1984:298, fig. 4) noted that the stratigraphic origin of that specimen was "highly uncertain" because USGS studies indicated only the presence of Holocene sediments overlying the Oligocene Cooper (now Ashley) Formation in the immediate area of the river at the tapir locality. Conceding that the tapir skull might be of Holocene age, Ray and Sanders (1984:298) thought it more probable that the specimen was washed out of Wando Formation deposits concealed by the Holocene tidal marsh deposits. Subsequent mapping by the USGS (Weems and Lemon, 1989, fence diagram A) has shown that at a point on the West Branch of the river approximately 1.2 km directly southeast of

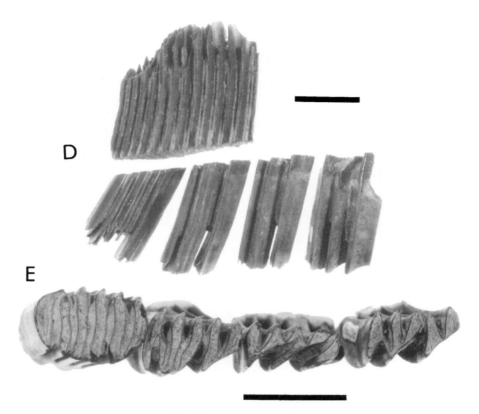

Figure 42 (cont.) **D**, right p4-m3, labial view, with right M3 in occlusion with m2 and m3. **E**, right p4-m3, occlusal view; Scale bars = 20 mm.

the *Tapirus* and *Neochoerus* locality the Wando Formation is indeed present below the thick deposit of Holocene sediments. It has been channeled out on the north side of the river at that point but does crop out as a thin exposure overlying the Ashley Formation at the bottom of the river. Thus, there is now little reason to doubt that the *Tapirus* and *Neochoerus* specimens from a short distance upriver are of late Pleistocene (Sangamonian) age.

The molar (ChM PV2439) and incisor (ChM PV2440) from Edisto Beach reported by Roth and Laerm (1980:14) place *N. pinckneyi* in the Wisconsinan (Late Rancholabrean) of South Carolina, and extend the temporal range of this species from Mid Pliocene to Late Pleistocene time. The genus is certainly not unknown in the Pliocene, the form *N. dichroplax* Ahearn and Lance, 1980, occurring in the Late Blancan of Arizona and Florida (Ahearn, 1981:62, 63). In Florida the genus is also represented by an undentified species from the Early Pleistocene Leisey Shell Pits, and *N. pinckneyi* is known from the latter part of the Early Pleistocene (Middle Irvingtonian) through the Late Pleistocene (Late Rancholabrean) (Morgan and White, 1995:453, table 7). None of the Charleston Museum teeth of *Neochoerus* have the bifurcated laminae characteristic of *N. dichroplax* teeth (Ahearn, 1981:65) and thus are safely referred to *N. pinckneyi*. The other described species of *Neochoerus*—*N.* magnus, *N. robustus*, *N. siraskae*, *N. sulcidens*, and *N. tarijensis*—are known only from Pleistocene beds in Central and South America (Ahearn, 1981:70).

The *Neochoerus pinckneyi* material from the Mid-Pliocene (3.8-3.6 Ma) Goose Creek Limestone appears to constitute the oldest record of capybaras in North, Central, or South America, but these animals almost certainly originated in tropical America, the ancestor of *N. pinckneyi* perhaps having moved northward along the Isthmian Link (Savage, 1974:44, fig. 3) and into North America during Early Pliocene (Early Blancan) time.

* * *

Order PERISSODACTYLA
Family EQUIDAE
Genus *Equus* Linnaeus, 1758
EQUUS SP.
Figures 43, 44, 45, 46

MATERIAL.—ChM PV5800, mandibular symphysis with three incisors; ChM PV5801, left m2, ChM PV5802, left M2; ChM PV5803, partial left P3; ChM PV5804, left tibia; ChM PV5831, right P2; ChM PV5832, left P3; ChM PV5833, left I1; ChM PV5834, right m1, ChM PV5848, right p2; ChM PV5836, left M1; ChM PV5840, left P2; ChM PV5779, proximal phalanx.

LOCALITY AND HORIZON.—ChM PV5800-PV5804: S.C., Charleston Co.; east bank of lake at Trailwood Trailer Park, 183 m southeast of Ree Street, east side of South Carolina Route 642 (Dorchester Road), c. 12 km northwest of Charleston; A.E. Sanders, P.S. Coleman et al.; associated with megathere remains (ChM PV4748) excavated November 1982; Penholoway Formation, Early Pleistocene. ChM PV5831-PV5834: S.C., Berkeley Co.; 68 m northwest of U.S. Route 52, 3.86 km southwest of old U.S. Route 52 in Moncks Corner (33° 10.1′ N., 81° 01.7′ W., U.S.G.S. Moncks Corner 7.5′ quadrangle); A.E. Sanders and party; associated with megathere remains (ChM PV4803) excavated May-June 1975; Ladson Formation, Middle Pleistocene. ChM PV5848: S.C., Berkeley Co.; west side of U.S. Route 176, 1.29 km north of U.S. Route 52 in Goose Creek; Bill Palmer, 20 July 1997; Ladson Formation, Middle Pleistocene. ChM PV5836: S.C., Dorchester Co.; ditch behind oxidation pond formerly located on north side of Ladson Road (County Road 230) and east side of Chandler Bridge Creek canal; Mike Crowell, summer 1978; Ten Mile Hill Beds, Late Middle Pleistocene. ChM PV5840: S.C., Dorchester Co.; spoil pile at O.K. Tire store, New Trolley Road (County Road 199); Vance McCollum, summer 1981; Ten Mile Hill Beds, Late Middle Pleistocene. ChM PV5779: S.C., Berkeley Co.; bank of Lindley Creek, 300 feet upstream from bridge 0.4 mi. east of S.C. Route 52; Billy Palmer, September 1991; Ten Mile Hill Beds, Late Middle Pleistocene.

AGE.—ChM PV5800, PV5801, PV5802, PV5804: Middle Irvingtonian, Pre-Illinoian. ChM PV5831, PV5832, PV5833, PV5834, and PV5848: Late Irvingtonian, Pre-Illinoian. ChM; ChM PV5779, PV5836 and PV5840: Early Rancholabrean, Late Illinoian.

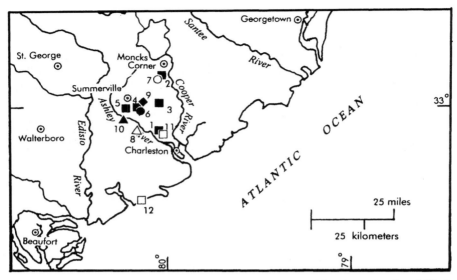

Figure 43. New locality records in South Carolina for *Equus* sp. (■ 1-5), *Tapirus haysii* (● 6), *Tapirus* cf. *T.* veroensis (○ 7), *Hemiauchenia* cf. *H. macrocephala* (△ 8), *Rangifer* cf. *R. tarandus* (◆ 9), *Cervalces scotti* (▲ 10), and *Cervus elaphus* (□ 11, 12). **1**, ChM PV5800-PV5804, Charleston Co., Trailwood Trailer Park, Penholoway Fm. (Early Pleistocene); **2**, ChM PV5831-5834, Berkeley Co., near Moncks Corner, Ladson Fm. (Middle Pleistocene); **3**, ChM PV5848, Berkeley Co., Goose Creek, Ladson Formation, (Middle Pleistocene). **4**, ChM PV5836, Dorchester Co., ditch on S. side of trailer park on N. side of Ladson Road (County Road 230) and E. side of Chandler Bridge Creek canal, Ten Mile Hill Beds (Late Middle Pleistocene); **5**, ChM PV5840, Dorchester Co., spoil pile onTrolley Road, Ten Mile Hill Beds (Late Middle Pleistocene). **6**, ChM PV5846, Dorchester Co.; ditch on N. side of trailer park, ca. 0.48 km (0.3 mi.) north of Ladson Road (Co. Rd. 230) and ca. 0.3 km (0.2 mi.) east of Chandler Bridge Creek canal; Ten Mile Hill Beds (Late Middle Pleistocene). **7**, ChM PV6022, Berkeley Co.; near Moncks Corner; Ladson Fm. (Middle Pleistocene). **8**, Berkeley Co., borrow pit for Mark Clark Expressway, E. side of S.C. Route 61; Wando Fm. (Late Pleistocene). **9**, ChM PV3411, Dorchester Co.; spoil pile, S. side of Chandler Bridge Creek and W side of County Road 377; Wando Fm. (Late Pleistocene). **10**, ChM PV2551, Charleston Co.; Magnolia Phosphate Mine, Runnymede Plantation; Wando Fm. (Late Pleistocene). **11**, ChM PV5805, Charleston Co.; Trailwood trailer Park; Penholoway Fm. (Early Pleistocene); **12**, ChM PV6076, Charleston Co., Edisto Beach; undetermined offshore bed (Late Pleistocene).

DISCUSSION.— Teeth and miscellaneous postcranial elements of Late Pleistocene horses of the genus *Equus* have been reported from South Carolina by Leidy (1860), Hay (1923a), Allen (1926), Roth and Laerm (1980), and Bentley, et al. (1995), but heretofore no specimens of early and middle Pleistocene age have been recorded from the state. Four specimens from the middle Irvingtonian, five from the late Irvingtonian, and two of early Rancholabrean age have entered the Charleston Museum collection within the past 20 years. In view of the somewhat unsettled state of equid taxonomy at the specific level, I have not attempted to make specific allocations.

The early Pleistocene (middle Irvingtonian) material (ChM PV5800-PV5804) from the Penholoway Formation was recovered during excavation of the partial

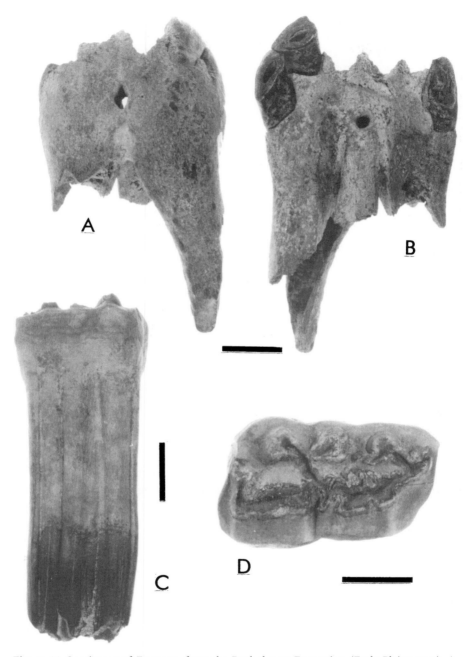

Figure 44. Specimens of *Equus* sp. from the Penholoway Formation (Early Pleistocene). **A**, mandibular symphysis (ChM PV5801) with left i2 and i3, roots of right i1 and i2, and partial i3; ventral view; **B**, dorsal view; scale bar = 20 mm. **C**, left m2 (ChM PV5801), labial view; scale bar = 15 mm; **D**, occlusal view; scale bar = 10 mm. S.C, Charleston Co., Trailwood Trailer Park.

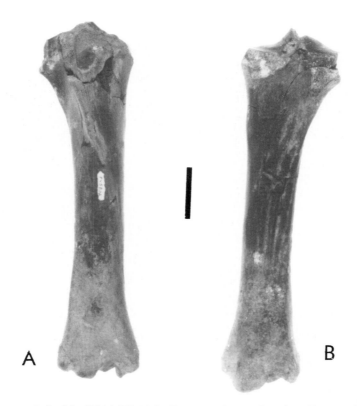

Figure 45. Left tibia (ChM PV5804), *Equus* sp. **A**, anterior view; **B**, posterior view. S.C, Charleston Co., Trailwood Trailer Park; Penholoway Formation (Early Pleistocene). Scale bar = 50 mm.

skeleton of *Eremotherium* sp. (ChM PV4748) discussed above. A well-preserved mandibular symphysis (ChM PV5800) (Figure 44a-b) contains the left second and third incisors, the right third incisor, and roots in alveolae of the right first and second incisors. This specimen is 111.5 mm in anteroposterior length as preserved, and is 65.2 mm in transverse diameter at the third inicsors. A left m2 (ChM PV5801, Figure 44c-d) and a small left P3 (ChM PV5803) also were found, the latter specimen closely resembling Leidy's (1860: pl.15) figure of his *Equus fraternus*. This material also includes a left tibia (Figure 45) measuring 350 mm in proximodistal length.

The middle Pleistocene (late Irvingtonian) specimens are five teeth, four of which—ChM PV5831 (Figures 46a-b), PV5832 (Figure 46c), PV5833, and PV5834 (Figures 46d-e)—were found in the Ladson Formation in association with the partial skeleton of *Eremotherium laurillardi* (ChM PV4803) reported above. Chm PV5848, a right p2, was collected from the Ladson Formation by Bill Palmer in July 1997.

The late middle Pleistocene (early Rancholabrean) specimens are from the Ten Mile Hill Beds north of Charleston and consist of a left M1 (ChM PV5836, Figures 46f-g), a left P2 (ChM PV5840), and a proximal phalanx (ChM PV5779).

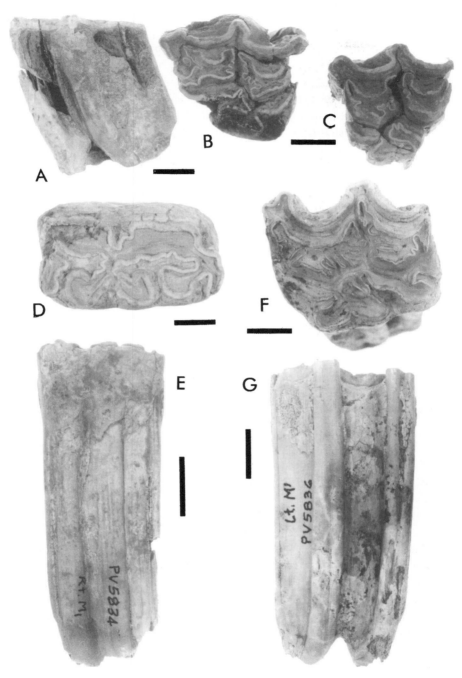

Figure 46. Teeth of *Equus* sp. from Middle Pleistocene deposits in South Carolina. **A**, partial right p2 (ChM PV5831), labial view; **B**, occlusal view; **C**, left P3 (ChM PV5832), occlusal view; **D**, right m1 (ChM PV5834), occlusal view; **E**, labial view; Berkeley County, Ladson Formation (Middle Pleistocene) **F**, left M1 (ChM PV5836), occlusal view; **G**, labial view; Dorchester County, Ten Mile Hill Beds (Late Middle Pleistocene). Scale bars = 10 mm.

Family TAPIRIDAE
Genus *Tapirus* Brisson, 1762
TAPIRUS HAYSII (LEIDY), 1859
Figures 43, 47

MATERIAL.—USNM 244383, right m3 (Ray and Sanders, 1984:295); USNM 347323, right m2 (Ray and Sanders, 1984:295); ChM PV5846, distal end of left tibia.

LOCALITY AND HORIZON.—USNM 244383, USNM 347323: S.C., Horry Co.; Myrtle Beach; Waccamaw Formation, Early Pleistocene. ChM PV5846: S.C., Dorchester Co.; ditch on north side of trailer park, ca. 0.48 km (0.3 mi.) north of Ladson Road (Co. Rd. 230) and ca. 0.3 km (0.2 mi.) east of Chandler Bridge Creek canal; A.E. Sanders, H.A. Stokes, C. Linder; summer 1990; Ten Mile Hill Beds, Late Middle Pleistocene.

AGE.—USNM 244383, USNM 347323: Early Irvingtonian, Pre-Illinoian. ChM PV5846: Early Rancholabrean, Late Illinoian.

DISCUSSION.—The name *Tapirus haysii* was applied by Leidy (1852a:106) to an isolated p4 (ANSP 11504) from North Carolina. A *nomen nudum* initially, it was used again by Leidy (1852b:106) for a partial mandible from Natchez, Mississippi, but again with no formal description of the taxon. Subsequently, he applied it to "an inferior back molar tooth" that "is larger than in the recent *Tapirus americanus*" (Leidy, 1854b:200), inferring for the first time the existence of an exceptionally large

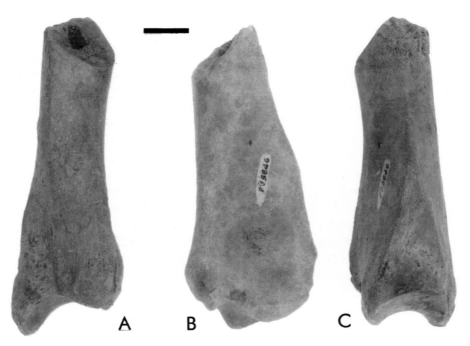

Figure 47. Distal end of left tibia, *Tapirus haysii* (ChM PV5846). **A**, posterior view; **B**, right lateral view; **C**, anterior view. S.C., Dorchester Co.; Ten Mile Hill Beds (Late Middle Pleistocene). Scale bar = 20 mm.

extinct tapir in North America. In his "Description of Vertebrate Fossils" in Holmes's *Post-Pleiocene Fossils of South Carolina*, Leidy (1859:106–107, pl. 17) included an account generally regarded as the valid type description of *T. haysii*, including the first figures and measurements of specimens of that taxon. Simpson (1945:66) questioned the validity of *T. haysii* and erected the new species *T. copei*, to which he referred *T. haysii*. Uncertainty about the type locality of *T. haysii* prompted Simpson (1945:65) to state that "the locality is not now known and can never be established with certainty." However, a splendid piece of research by Clayton E. Ray later provided evidence that the type of *T. haysii* was collected by Thomas Nuttall on the north bank of the Neuse River, 16 miles below New Bern, North Carolina, in 1832 (Ray and Sanders, 1984:288). With that question resolved, Ray and Sanders (1984:297) revived the name *T. haysii* and considered *T. copei* to be a junior synonym of that. taxon, noting that "*T. copei* could be faulted by standards similar to those on which *T. haysii* was rejected, because skull characters are not available. It would be convenient to have better type specimens for many paleotaxa, but stability is served best by conserving named taxa if at all possible." Subsequently, 12 new records from Florida provided the first cranial material of *T. haysii* (Hulbert, 1995), placing this taxon on a much sounder taxonomic footing. Hulbert (1995) found great similarity in the cranial morphology of *T. haysii* and *T. veroensis*, and in a recent phylogenetic analysis (Hulbert, 1999) he placed *T. haysii* in a clade with *T. veroensis* and *T. merriami*. The best known and most frequently encountered tapir in the fossil record of North America, the smaller *T. veroensis* has been reported from Late Pleistocene beds from New York to Florida and westward to Tennessee and Illinois. *T. haysii* has been recorded from Late Pliocene to Middle Pleistocene deposits from Pennsylvania to Florida and westward to Nebraska, Colorado, and Arizona (Ray and Sanders, 1984, fig. 3; Hulbert, 1995:520). The Irvingtonian *T. merriami*, also a relatively large form, was regarded as "probably conspecific" with *T. haysii* by Ray and Sanders (1984), but Jefferson (1989) recognized specimens from California and Arizona as valid *T. merriami*. The taxonomic status of *T. merriami* must, however, remain in question until the type skull from California can be compared with the cranial material of *T. haysii* from Florida.

Ray and Sanders (1984:295) reported only two specimens from South Carolina that "more or less certainly" can be assigned to *T. haysii*, viz., a right m2 (USNM 347323) and a right m3 (USNM 244383), both from Myrtle Beach, Horry County. Since then, another specimen clearly assignable to *T. haysii* has come to light. During the summer of 1990 the distal end of the left tibia (ChM PV5846, Figure 43) of a large tapir was found in the Ten Mile Hill Beds (Late Middle Pleistocene) in the side of a drainage ditch approximately 0.48 km north of Ladson Road (County Road 230) in Dorchester County, ca. 20 miles north of Charleston by Aaron Stokes, Chester Linder, and the writer. Measurements (in mm) of ChM PV5846 are as follows:

Proximodistal length, as preserved	157.3
Width of articular surface	48.8
Anteroposterior length, articular surface	37.8

The measurements of the width and length of the distal articular face of this specimen exceed those of the largest tibia (46.3 mm wide, 36.6 mm long) of *T. haysii* examined by Hulbert (1995:540, table 4), indicating an individual of considerable size.

As reflected in Table 2, the present study does not report any *T. haysii* specimens from South Carolina younger than the late middle Pleistocene. The age of the two teeth from Myrtle Beach (USNM 347323 and 244383) is uncertain, but it is probable that they came from the early Pleistocene Waccamaw Formation rather than the Socastee, considering that *T. haysii* has not yet been recorded from the Late Pleistocene.

* * *

TAPIRUS CF. T. VEROENSIS SELLARDS, 1918
Figures 43, 48

MATERIAL.—ChM PV6022, anterior thoracic vertebra.

LOCALITY AND HORIZON.—S.C., Berkeley Co.; 68 m northwest of U.S. Route 52, 3.86 km southwest of old U.S. Route 52 in Moncks Corner (33° 10.1′ N., 81° 01.7' W., USGS Moncks Corner 7.5′ quadrangle); A.E. Sanders and party, May-June 1975; Ladson Formation, Middle Pleistocene.

AGE.—Late Irvingtonian, Pre-Illinoian.

DISCUSSION—In their review of Pleistocene tapirs in the eastern United States, Ray and Sanders (1984:312) noted that "The available specimens indicate that *T. veroensis* was very widespread in the conterminous United States in late Pleistocene time, at least from Texas and Missouri eastward through Tennessee to South Carolina; that it may have extended from coast to coast and north to within the glaciated northern states, and possibly back to earlier Pleistocene time, if more fragmentary specimens are correctly referred."

One recently-determined specimen seems to provide evidence that *T. veroensis* was present in South Carolina during the middle Pleistocene. This specimen, an anterior thoracic vertebra (ChM PV6022, Figure 48), was found in association with a partial skeleton of *Eremotherium laurillardi* (ChM PV4803) excavated from the Ladson Formation by The Charleston Museum in 1975 (see discussion above). It lay unidentified among material from the excavation until recently, when it was identified by Clayton E. Ray as a thoracic vertebra of *Tapirus,* possibly the fourth in the thoracic series. Although most of the posterior epiphysis has been broken away from the centrum, the portion that remains is firmly ankylosed to the centrum, as is the complete anterior epiphysis, leaving no doubt that the specimen came from an adult individual. It is too small to be assigned to *Tapirus haysii,* the large Pleistocene tapir of North America, and thus appears referable to *T. veroensis* (C. Ray, personal communication, August 2000).

Measurements (in mm) of ChM PV6022 are as follows:

Anteroposterior length of centrum	34.2
Transverse diameter, anterior end of centrum	33.3
Transverse diameter, posterior end of centrum	59.9
Height of neural arch, anteriorly	13.1

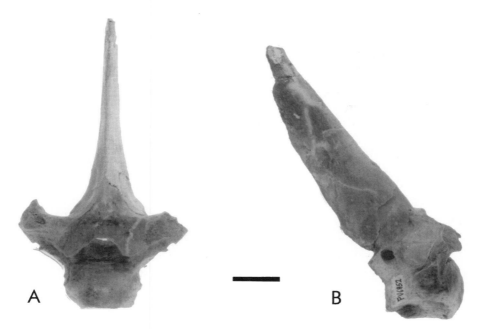

Figure 48. Anterior thoracic vertebra, *Tapirus* cf. *T. veroensis.* **A**, anterior view; **B**, right lateral view. S.C., Berkeley Co., Ladson Fm. (Middle Pleistocene). Scale bar = 20 mm.

Height of neural arch, posteriorly	14.4
Ventral face of centrum to tip of spinous process, as preserved	143.7
Length of centrum into transverse width of anterior end	0.97

If this specimen is indeed assignable to *T. veroensis* it would extend the temporal range of that species well back into late Irvingtonian time.

* * *

Order ARTIODACTYLA
Family CAMELIDAE
Subfamily CAMELINAE
Tribe Lamini
Genus *Hemiauchenia* H. Gervais and Ameghimo, 1880
HEMIAUCHENIA CF. *H. MACROCEPHALA* (COPE, 1893)
Figures 43, 49a-b

MATERIAL.—Proximal end of phalanx (ChM PV5854).

LOCALITY AND HORIZON.—S.C., Charleston Co.; spoil pile at borrow pit for Mark Clark Expressway, E. side of S.C. Route 61 near Ashley Hall Plantation Road; Doris Holt, December 1979; Wando Formation, Late Pleistocene.

AGE.—Rancholabrean, Sangamonian.

DISCUSSION.—Camelids were first reported from South Carolina by Hay (1923a:363) who listed "Camelops sp. indet." among the Pleistocene taxa represented

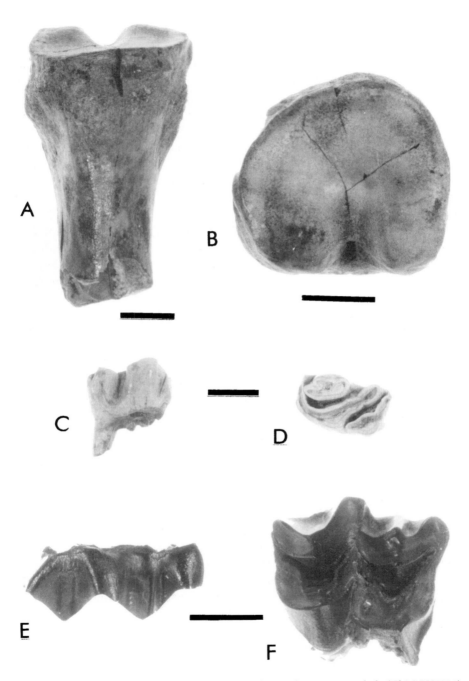

Figure 49. A, proximal end of phalanx, *Hemiauchenia* cf. *H. macrocephala* (ChM PV5854), dorsal view; **B**, articular face. **C**, left p4, *Rangifer* cf. *R. tarandus* (ChM PV3411), lingual view; **D**, occlusal view. **E**, crown of right DP4, holotype of *Alces runnymedensis* Hay, 1916 (ChM PV2551), herein referred to *Cervalces scotti*, labial view; **F**, occlusal view. Wando Fm. (Late Pleistocene). Scale bars = 10 mm.

in specimens that he examined in Charleston in 1915. The specimen upon which the record is based was not specified, but it is quite doubtful that it belonged to the genus *Camelops* in view of Webb's (1974:207) discovery that all of the Florida specimens assigned to *Camelops* were actually referable to *Palaeolama mirifica* (Simpson, 1929). His conclusion that it was unlikely that *Camelops* ever occurred in eastern North America was verified in Kurtén and Anderson's (1980) distribution map for *Camelops*, which shows that this genus has not been validly recorded east of the Mississippi River.

"A well-preserved astragalus of a camel from Ashley River" found during the phosphate mining days was referred to *Procamelus minor* (Leidy) by Allen (1926:452) based on the similarity of its measurements to a specimen of *P. medius* from Florida. That assignment is highly improbable because the temporal range of *Procamelus* was restricted to the Miocene (McKenna and Bell, 1997), and all of the Miocene sediments have been eroded on the northeastern half of coastal South Carolina from Charleston to the Cape Fear Arch except beneath the barrier islands near Charleston, where the early Miocene Marks Head Formation unconformably underlies the Wando Formation (Figure 2). Re-examination of this MCZ specimen probably will show it to be referable to *Palaeolama* or *Hemiauchenia*.

Roth and Laerm (1980) reported four camelid limb elements in the ChM collections (ChM PV2407, PV2409, PV2410, PV2411) as *Palaeolama* cf. *P. mirifica*. The morphology of a well-preserved right radius-ulna with carpals in place (PV2409) is in basic agreement with Webb's (1974c, fig, 9.7B) illustration of this bone in *P. mirifica*. The other three specimens may warrant re-evaluation in view of the very recent discovery that the large llama *Hemiauchenia* also occurred in South Carolina during Pleistocene time.

In 1979 the proximal end of a phalanx displaying camelid affinities was found by a Charleston Museum volunteer on a spoil pile of Wando Formation sediments excavated from a borrow pit along S.C. Rt. 61. This specimen (ChM PV5854, Figures 49a-b) remained undetermined until May 2002 when it was sent to David Webb (Florida Museum of Natural History), who identified it as *Hemiauchenia* and probably referable to *H. macrocephala* (S.D. Webb, personal communication, 28 May 2002). As preserved, this partial phalanx measures 59.15 mm proximodistally and 31.5 mm in greatest transverse diameter. Although *Hemiachenia* occured in North America from the Middle Pliocene to the Late Pleistocene (Webb, 1974c), this specimen is the first evidence of the genus in South Carolina and extends its range northward along the Atlantic coast from Florida, where it has been recorded from numerous localities (Kurtén and Anderson, 1980, fig. 15.7)

Llamas originated in North America during the late Tertiary, and during the early Pleistocene *Hemiauchenia* extended its range well into South America, the first of the lamines to reach that continent (Webb, 1974c). Kurtén and Anderson (1980 fig. 15.7) show two species of *Hemiauchenia* occuring in the western United States during Pleistocene time, *H. blancoensis* and *H. macrocephala*, both of which have been recorded from Florida. Lamine remains from numerous other localities in the West are mapped by those authors as *Hemiauchenia* sp. and are in need of specific resolution.

Family CERVIDAE
Subfamily ODOCOILEINAE
Genus *Rangifer* Hamilton-Smith, 1827
RANGIFER CF. R. TARANDUS (LINNAEUS, 1758)
Figures 43, 49c-d

MATERIAL.—Lower left fourth premolar (ChM PV3411).

LOCALITY AND HORIZON.—S.C., Dorchester Co.; spoil pile on south side of Chandler Bridge Creek and west side of County Road 377 (Jamison Road); Vance McCollum, February 1981. Wando Formation, Late Pleistocene.

AGE.—Rancholabrean, Sangamonian.

DISCUSSION.—This specimen was briefly noted by McDonald et al. (1996:426) in an addendum to their discussion of the Pleistocene distribution of caribou in the eastern United States. The tooth was stated to have come from "the Chandler Bridge locality" in Dorchester County, but inasmuch as Chandler Bridge, located on County Road 230, is more than two kilometers southeast of where the specimen was actually collected, it seems useful to place the correct locality on record and to provide stratigraphic data.

The tooth, a well-preserved left p4 (ChM PV3411, Figures 46c-d), was found by Vance McCollum in February 1981 on a spoil pile in a cleared area bordered by Chandler Bridge Creek on the north and by County Road 377 (Jamison Road) on the east, this location being approximately 1.8 miles (2.9 km) northwest of County Road 230 (Ladson Road) (USGS Stallsville 7.5′ quadrangle). The spoil pile consisted primarily of sediments excavated from the Wando Formation, which in that area overlies the Upper Oligocene Chandler Bridge Formation. The specimen was donated to The Charleston Museum and later was tentatively identified as *Rangifer* by Clayton E. Ray (National Museum of Natural History). McDonald et al. (1996:426) noted that the tooth has little wear and "can be matched closely by those of large male individuals of *Rangifer tarandus* in the collections of the U.S. National Museum." McDonald et al. (1996:422) reported a partial antler (USNM 467795) from the Intracoastal Waterway near Myrtle Beach, Horry County, that in their opinion "most nearly resembles the antlers of *Rangifer*," but allowing "that this specimen might eventually be shown to belong to another taxon" they preferred to consider it as ?*Rangifer*.

Churcher et al. (1989) recorded caribou from the late Pleistocene of northwestern Alabama, and McDonald et al. (1996:409, fig. 1) documented important new records of *Rangifer* from northern Mississippi, the Virginia-North Carolina border, the Continental Shelf 13 miles off Currituck Beach, North Carolina, New Bern, North Carolina, and from the two South Carolina localities noted above. These records extended the Pleistocene distribution of *Rangifer* onto the Gulf and Atlantic coastal plains, considerably further south than it had been known previously.

McDonald et al. (1996:424) observed that the distribution pattern of *Rangifer* in the eastern United States reflects the shape of the Wisconsinan Laurentide glacial system, and that, plus "the fact that *Rangifer* is a boreal ungulate, strongly suggests that *Rangifer* probably reached its maximum southeasterly distribution during glacial time, an idea that has been accepted for many years. The fact that three of the specimens described . . . came from the Continental Shelf or the surf along the Atlantic Coast indicates that *Rangifer* occupied the region during some glacial stage(s) when sea level was lower and the Continental Shelf was emergent." ChM PV3411, from the Wando Formation of Sangamonian age, suggests, however, that *Rangifer* occurred on the South Atlantic Coastal Plain during warmer periods as well.

* * *

Genus *Cervalces* Scott, 1885
CERVALCES SCOTTI (LYDEKKER, 1898)
Figures 43, 49e-f

MATERIAL.—ChM PV2551: The crown of a deciduous upper right fourth premolar, the holotype of *Alces runneymedensis* Hay, 1923.

LOCALITY AND HORIZON.—S.C., Charleston Co.; Magnolia Phosphate Mine, Runnymede Plantation, ca. 17.7 km (11.0 mi.) northwest of Charleston; Charles C. Pinckney, c. 1900; Wando Formation, Late Pleistocene.

AGE—Rancholabrean, Sangamonian.

DISCUSSION.—"*Alces runnymedensis* was first briefly referred to in Year Book No. 14 of the Carnegie Institution of Washington, 1915 (1916), page 387" (Hay, 1923a:364). A right DP4 missing the roots, it was in the private collection of Charles C. Pinckney, owner and operator of the Magnolia Phosphate mine, at the time that Hay examined it and described it as a new species of moose, *A. runnymedensis*, the specific name being derived from the name of Pinckney's estate, Runnymede Plantation. Identifying it as "an upper right hindermost milk molar," Hay (1923a:364) noted that "The tooth closely resembles the corresponding one of *Alces americanus* [= *Alces alces*], but is larger and has a flatter crown. Only the crown of the tooth is preserved, and of this a part of the enamel of the inner anterior cone is broken off; otherwise it is in fine condition. The color is very black."

Kurtén and Anderson (1980:316) synonymized this taxon with *Alces alces* without explanation and without examining the holotype; but recent unpublished findings by Dale Guthrie (personal comunication, July 1997) indicate that the genus *Alces* is a dubious place for this specimen. Based upon C^{14} dates obtained by Guthrie from North American fossil specimens of *Alces*, representatives of that genus did not enter North America before about 12,000 years ago. Like other specimens in the Pinckney collection, the black coloration of this tooth indicates that its most likely origin was the lag deposit at the base of the Wando Formation, which, as previously noted, is about 100,000 years old in the phosphate mining area along the Ashley River, far too early to have been a repository for moose remains within the temporal constraints observed by Guthrie (personal communiation, July 1997).

Based upon present knowledge of the occurrence of large, moose-like cervids in the Pleistocene of the eastern United States, the stag-moose, *Cervalces scotti*, is the only known form that might have been present on the Coastal Plain of South Carolina during Middle Wando time. ChM PV2551 was compared with the dentition in a cast of a fragment of a left maxilla (ISM 494014) of *C. scotti* from late Pleistocene bog sediments near Kendallville, Noble, Co., Indiana. The original specimen of the latter is in a private collection but was measured and figured by Farlow and McClain (1996:327, fig. 3.). P2–M3 are preserved and show considerable wear, indicating an animal well into maturity, as is suggested also by the large, massively-built antler rack found with the maxillary fragment (Farlow and McClain, 1996:327, fig. 2.). Allowing for the worn occlusal surfaces of the teeth in ISM 494014 and the differences between permanent and deciduous dentition, ChM PV2551 compares quite favorably with the P4 of the former specimen in both size and morphology. The chief difference between them is that the division of the lingual margin into two cones is more strongly developed in ChM PV2551, but the cast (ISM 494014) of the Indiana specimen shows that the posterior cone is broken off in that specimen, as it is in ChM PV2551. Comparative measurements (in mm) of the holotype DP4 of *Alces runnymedensis* (ChM PV2551) and a P4 of "*Alces americanus*" (= *Alces alces*) (USNM 117055) as given by Hay (1923a:364), and those of the original P4 in the maxillary fragment of *C. scotti* as recorded by Farlow and McClain (1996:327) are as follows, with Hay's wording in quotation marks:

	P4 *Alces alces* USNM 117055	DP4 *A. runnymedensis* ChM PV2551	P4 *C. scotti*
Anteroposterior length of occlusal surface	24.0	25.5	25
Occlusal width (= "Width of tooth from median style to base of inner hinder cone")	21.0	24.0	24
"Length of tooth at middle width"	21.5	23.0	—
"Width of tooth along front border"	23.0	23.0	—

As shown above, the occlusal widths and lengths of ChM PV2551 and the original P4 of ISM 494014 are almost identical, despite the fact that the former specimen is a deciduous tooth.

Thus, on both temporal and morphometric grounds, the holotype DP4 of *A. runnymedensis* Hay, 1923 (ChM PV2551) now seems referable to *Cervalces* and is herein assigned to *Cervalces scotti* (Lydekker), 1898, the only known species in the genus.

Churcher's (1991:394, fig. 2) distribution map for *Cervalces* shows it to have occurred as far south as extreme southwestern Virginia, not an unreasonable distance from the Lower Coastal Plain of South Carolina, which now appears to be the southernmost record of *C. scotti*.

Subfamily CERVINAE
Genus *Cervus* Linnaeus, 1758
CERVUS ELAPHUS LINNAEUS, 1758
Figures 43, 50

MATERIAL.— ChM PV5805, first lumbar vertebra; ChM PV6076, lower right third molar.

LOCALITY AND HORIZON.—ChM PV5805: S.C., Charleston Co.; east bank of lake at Trailwood Trailer Park, 183 m southeast of Ree Street, east side of South Carolina Route 642 (Dorchester Road), c. 12 km northwest of Charleston; A.E. Sanders, P.S. Coleman et al.; associated with megathere remains (ChM PV4748) excavated November 1982; Penholoway Formation, Early Pleistocene. ChM PV6076: S.C., Charleston Co.; Edisto Beach; Doris Holt; undetermined offshore beds, Late Pleistocene.

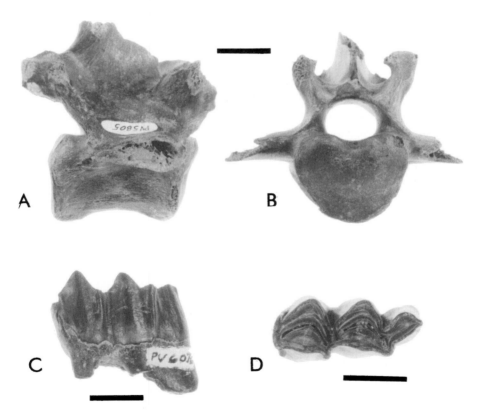

Figure 50. Specimens of *Cervus elaphus* from the Pleistocene of South Carolina. **A**, first lumbar vertebra (ChM PV5805), right lateral view; **B**, anterior view; Charleston Co.; Penholoway Fm. (Early Pleistocene); scale bar = 20 mm. **C**, right m3 (ChM PV6076), labial view; **D**, occlusal view; Charleston Co.; undetermined offshore beds (Late Pleistocene); scale bars = 10 mm.

AGE.—ChM PV5805: Irvingtonian, Pre-Illinoian. ChM PV6076: Rancholabrean, Late Wisconsinan.

DISCUSSION.—In North America elk are known from the Irvingtonian Cape Deceit fauna of Alaska (Kurtén and Anderson, 1980:318) and survive to the present day. "*C. canadensis*" (= *C. elaphus*) was reported from the Pleistocene of South Carolina by Hay (1923a:242, 363) on the basis of two teeth (ANSP 11565, 11566) collected by Captain A.H. Bowman along the Ashley River. "It is possible that Leidy [1860] did not mention them because he regarded them as teeth of elk that lived within Recent times" (Hay, 1923a:242). More recently, a right m3 of *C. elaphus* and a cervid vertebra that appears to be referable to that taxon have been found.

The right m3 (ChM PV6076, Figure 50c-d), collected on Edisto Beach by Doris Holt, compares favorably with specimens of *C. elaphus* dentition in the Charleston Museum. This tooth is the first definite record of *C. elaphus* among the Rancholabrean mammal remains from Edisto Beach.

One cervid vertebra (ChM PV5805, Figure 50a-b) was among the faunal remains found in association with the eremothere remains (ChM PV4748) recovered from the Penholoway Formation (Early Pleistocene), discussed above as *Eremotherium* sp. Though missing its spinous process and both transverse processes, the vertebra is well preserved otherwise. Comparison with a Recent skeleton of *C. elaphus* (ChM CM540) indicates that the specimen is a first lumbar veterbra. The centrum of this vertebra measures 53.2 mm in anteroposterior length and 42.9 mm in transverse diameter across its anterior face, yielding a length/width ratio of 0.81. There are slight differences in the morphology of the neural arch of the fossil and that of the first lumbar of the Recent specimen, but the similarities are sufficient to warrant assignment of ChM PV5805 to *C. elaphus* pending examination of additional Recent material demonstrating the degree of variation in first lumbars of this species.

* * *

Family BOVIDAE
Genus *Bison* (Smith, 1827)
BISON ANTIQUUS ANTIQUUS LEIDY, 1852
Figures 51, 52a-b, d

MATERIAL.—ChM PV2393, left horn core of mature male; ChM PV7000, left horn core of mature female.

LOCALITY AND HORIZON.— S.C., Charleston Co.; Edisto Beach; undetermined offshore beds, Late Pleistocene. ChM PV2393, Robert H. Coleman, 20 February 1947; ChM PV7000, Howard and Braith Eldridge, September 1995.

AGE— Rancholabrean, Late Wisconsinan.

DISCUSSION—ChM PV2393 (Figures 52a-b) was reported by Roth and Laerm (1980) as *Bison* cf. *B. antiquus* on the basis of horn core measurements in Skinner and Kaisen (1947). Following his examination of the Charleston Museum *Bison* material, Jerry McDonald noted that ChM PV2393 "does not conform exactly . . . with the horn core character complex of *Bison antiquus antiquus*, but it matches that taxon more

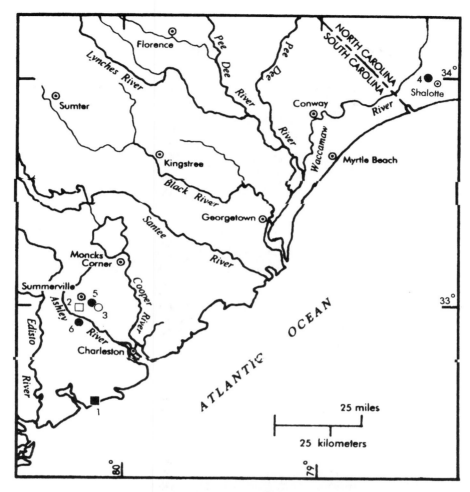

Figure 51. Localities for *Bison antiquuus antiquus* (■ 1), *Bison* sp. (□ 2), *Mammut americanum* (○ 3), and *Cuvieronius* sp. (● 4-6) in North Carolina and South Carolina. **1**, ChM PV2393, PV7000, S.C., Charleston Co., Edisto Beach; undetermined offshore beds (Late Pleistocene). **2**, ChM PV6865, S.C., Dorchester Co.; Irongate subdivision; Ten Mile Hill Beds (Late Middle Pleistocene). **3**, ChM PV6077, S.C., Dorchester Co.; Tranquil Estates subdivision; Wando Fm. (Late Pleistocene). **4**, USNM 391982, N.C., Brunswick Co.; pit near Shalotte; Waccamaw Fm. (Early Pleistocene); **5**, ChM PV4885, S.C., Dorchester Co., drainage ditch behind trailer park ca. 0.30 mi. (0.48 km) north of Ladson Road (County Road 230), ca. 0..20 mi. (0.32 km) east of Chandler Bridge Creek; Ten Mile Hill Beds (Late Middle Pleistocene). **6**, ChM PV2611, PV2652, S.C., Charleston Co., Magnolia Phosphate Mine, S.C. Rt. 61 near Runnymede Plantation; Wando Fm. (Late Pleistocene).

closely than any other. This is expected, based on the peripheral geographic location of the specimen's provenience. In addition, I have noticed that deviations from the 'ideal' morphology are substantially more frequent in what appears to have been peripheral populations than in the core populations" (personal communication, 15 June 2001). ChM PV2393 is badly eroded ventrally but is well preserved dorsally (Figure 52a).

A recently acquired horn core (ChM PV7000, Figure 52d) seems clearly referable to

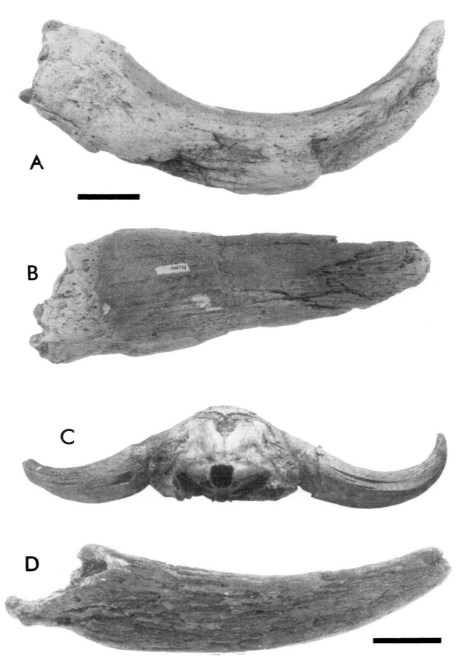

Figure 52. Specimens of *Bison antiquuus antiquus*. **A**, left horn core, mature male, (ChM PV2393), anterior view; **B**, dorsal view; S.C., Charleston Co.; Edisto Beach; undetermined off-shore unit (Late Pleistocene); scale bar = 50 mm. **C**, posterior view of skull from Aucilla River, Florida; late Pleistocene (from Hulbert, 2001, fig. 13.48B). **D**, left horn core, mature female, (ChM PV2393), anterior view; **E**, dorsal view; S.C., Charleston Co.; Edisto Beach; undetermined offshore unit (Late Pleistocene); scale bar = 40 mm.

B. a. antiquus. Collected on Edisto Beach in September 1995 by Howard and Braith El-dridge, this specimen is virtually complete and is missing only the very tip of the distal end and the dorsal potion of the base, the latter condition precluding any knowledge of the presence or absence of a burr. The specimen is considerably smaller than ChM PV2393 but shares the same basic configuration in anterior view (Figures 52a, d), i.e., a long sweeping upward curvature, not straight as in *B. latifrons* or abruptly curved as in *B. bison.* The degree of curvature in ChM PV7000 is more gradual than in the larger PV2393, and it is not as sharply upturned at the tip—a characrteristic of female horn cores in *B. antiquus* (J. McDonald, personal communication, 18 June 2002).

The following measurements (in mm) of the two specimens were taken using McDonald's (1981:fig. 11; table 21) morphometric characters:

	ChM PV2393	**ChM PV7000**
Horn core length, upper curve, tip to burr	330	245[+]
Straight line distance, tip to burr, dorsal horn core	285	—
Dorso-ventral diameter, horn core base	—	69 [+]

The measurements of PV2393 fall within the upper size limits of male *B. a. antiquus* horn cores reported in McDonald's (1981:77) table 21. Measurements of 32 horn cores along the upper curve ranged from 203-364 mm; those of 30 cores taken on a straight line from tip to burr ranged from 185-330 mm. ChM PV2393 is 91% of the maximum in the former measurement and 86% of the maximum in the latter.

The smaller PV7000 scores within McDonald's (1981:77, table 21) range of meas-urements for horn cores of female *B. a. antiquus,* viz., 145-253 mm for the length along the upper curve and 53-79 mm for the dorso-ventral diameter of the horn core base, and thus appears to be that of a mature female. The latter measurement was taken on the distal edge of the broken and missing dorsal portion of the core but it is stll a valid measure of its vertical diameter (McDonald, 1981: 45, fig. 7) and is well within McDonald's (1981) parameters for female horn cores. Those measure-ments and the flatter horizontal arc of curvature leave little doubt that ChM PV7000 is a horn core of a mature female *B. antiquus.*

In what is to date the most comprehensive study of North American *Bison,* Mc-Donald (1981) recognized five species and four subspecies that have occurred over various portions of the continent during Pleistocene and Holocene times. Those in-clude the modern *Bison bison bison* (Linnaeus) and *Bison bison athabascae* Rhoads, 1897, and five fossil forms: *Bison latifrons* (Harlan, 1825); *Bison antiquus antiquus* Leidy, 1852; *Bison antiquus occidentalis* (Lucas, 1898); and the Eurasian *Bison priscus* (Bojanus, 1827) and *Bison alaskensis* Rhoads, 1897, both of which ranged through Beringia into western North America and as far south as central Mexico. Other tax-onomic arrangements have been employed (e.g., Wilson, 1974), but McDonald's (1981) version is based upon the most thorough morphometric analysis of *Bison* yet attempted, and, as noted by Pinsof (1991:510), "[it] remains the most utilitarian and comprehensive classification available."

Compared with *B. latifrons*, "*B. a. antiquus* is the smaller, shorter horned bison of late Pleistocene-early and middle Holocene North America" (McDonald, 1981:77). McDonald (1981, figs.20, 21) shows *B. a. antiquus* to have been the most widely distributed of all North American bison, ranging from Alaska through western and central Canada and the United States southward through Mexico to Nicaragua, with an outlying population in Florida. In the subspecies *B. antiquus occidentalis* "The horn core tip is much thinner and more pointed than that of *B. a. antiquus*" (McDonald, 1981:86). *B. antiquus occidentalis* populations were centered on the Great Plains of Canada and the United States and the western prairies of the northern midwest (McDonald, 1981:92).

Bentley et al. (1995) referred several teeth and postcranial elements in their Ardis Local Fauna to *B. antiquus* on the basis of size. Roth and Laerm (1980) also referred a large atlas vertebra (ChM GPV2) from Edisto Beach to *Bison* cf. *B. antiquus* because of its size. The transverse diameter of its condylar facet (124 mm) does indeed fall within the range of McDonald's (1981, table 21) measurements of the width of the occipital condyles of female *B. a. antiquus* specimens (116-149 mm). On temporal and ecological grounds, most, if not all, of the *Bison* material from Edisto Beach is probably referable to *B. a. antiquus*. *B. bison* did not appear until late Holocene time (5,000-4,000 ybp) and *B. latifrons* seems to have preferred forested areas (McDonald, 1981) instead of the savanna-type habitat that apparently predominated over much of the now-submerged seaward portion of the Coastal Plain during its emergence in Wisconsinan time (see Summary and Conclusions).

The present records of *B. a. antiquus* (ChM PV2393, PV7000) verify the presence of this species in South Carolina, extend the range northward from Florida along the Atlantic coast, and suggest that this form may have occurred throughout the southeastern United States.

* * *

BISON SP.
Figures 51, 53

MATERIAL.—ChM PV6865, right astragulus

LOCALITY AND HORIZON.—S.C., Dorchester Co.; ditch in Irongate subdivision; Vance McCollum, 10 May 1981; Ten Mile Hill Beds, Late Middle Pleistocene.

AGE—Early Rancholabrean, Illinoian

DISCUSSION—Typical of fossil bones from the Ten Mile Hill Beds, this specimen is light biege to grayish white in color with areas of black and orange iron staining. Its morphological details agree completely with those of other *Bison* astragali in the ChM collection. ChM PV6865 is 87 mm in greatest proximodistal length and 50 mm in greatest anteroposterior length. The age of the beds in which it was found suggests that this specimen may be referable to *B. latifrons* based on temporal distributional models that favor sequential (vs. concurrent) species of *Bison*, but like most isolated postcranial elements of *Bison*, it cannot be assigned to any species with certainty.

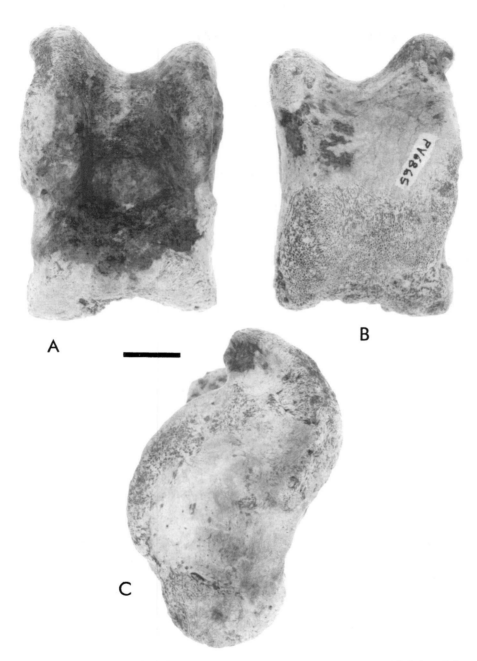

Figure 53. Right astragalus (ChM PV6865), *Bison* sp. **A**, anterior view; **B**, medial view. S.C., Dorchester Co.; Ten Mile Hill Beds (Late Middle Pleistocene). Scale bar = 15 mm. Specimen upon which Irvingtonian/Rancholabrean boundary (0.24 Ma) is founded in text.

By virtue of its discovery in the late middle Pleistocene Ten Mile Hill Beds, this specimen establishes the presence of bison on the east coast of North America between 0.24 and 0.20 Ma, the estimated period of deposition of the Ten mile Hill Beds based on coral dates by Szabo (1985). The late middle Pleistocene date for that unit is supported by the presence of *Tapirus haysii*, reported from the Ten Mile Hill beds in the present paper. *T. haysii* is not known from beds of Late Pleistocene age (Ray and Sanders, 1982).

The age of this specimen is of particular significance because it places bison on the east coast earlier than the date assigned to *Bison* material from the American Falls Formation along the Snake River near American Falls, Idaho, presently regarded as the earliest record of bison in the Pleistocene of North America (Ernest Lundelius, personal communication, April, 2002). Pinsov (1991) cited the age of that material as between 210 ± 60 ka and 72 ± 14 ka.

These dates are important because the boundary of the Irvingtonian and Racholabrean Land Mammal ages has been based upon the first appearance of *Bison* in North America (Savage, 1951; Kurtén and Anderson, 1980; Lundelius et al., 1987); but as noted by Lundelius et al. (1987:223), "the beginning of the Rancholabrean Mammal Age is poorly dated, with estimates ranging from 0.2 to 0.55 Ma." In their correlation of Plio-Pleistocene faunas, Lundelius et al. (1987:222, :fig.7.3) placed the Irvingtonian/Rancholabrean boundary at 0.3 Ma. This date was also used for the base of the Rancholabrean by Morgan and Hulbert (1995:20), "with the expectation that future work will more precisely document the arrival of *Bison* in North America."

The availability of a reasonably well-founded early date for *Bison* on the eastern seaboard provides an opportunity to place the boundary for the Irvingtonian/Rancholabrean Land Mammal ages with somewhat better precision than in the past, so it seems practical to propose the estimated age of the base of the Ten Mile Hill Beds—0.24 Ma—as the boundary of these two faunal stages. There is no apparent means of pinning the Ten Mile Hill Bed specimen (ChM PV6865) to a date within the time of deposition of that unit (0.24-0.20 Ma), so it seems safest to allow maximum latitude in its use as a boundary marker. The 0.24 Ma date coincides with the period of Illinoian glaciation, at which time the dispersal of *Bison* and several other Eurasian taxa through Beringia and into the conterminous United States seems to have taken place, "presumably at one of the climaxes of Illinoian glaciation" (Repenning, 1998:73).

There is, however, a recently published report of bison in Florida in Blancan time. McDonald and Morgan (1999) described a fragment of a horn core (UF 100486) from a bed in the Macasphalt Shell Pit in Sarasota County dated at ca. 2.5-1.9 Ma, and another horn core fragment (UF 18286) from the Inglis 1A site in Citrus County, the fauna of which has been dated at 1.9-1.6 Ma. Based on previous records, one would not expect the presence of *Bison* in Florida at such an early period. If a mean of 1.75 Ma is used for the 1.9–1.6 milliion-year date for the age of the Inglis 1A fauna, there is a difference of 1.51 million years between that date and that of the South Carolina specimen (ChM PV6865). Within that span of time there are no further records of bison known from Florida or anywhere else in the conterminous United States, although they are well represented in Alaska during that period, as indicated by revised dates of Pleistocene tephras by Péwé, et al. (1989). The genus is not present in the splendid early Pleistocene mammalian fauna from the Leisey Shell Pits (Morgan, and Hulbert, 1995) and is not recorded in the Pleistocene of Florida until it appears in the form of *Bison latifrons* in the Rancholabrean Haile 8A fauna, which was listed as Sangamonian (late Rancholabrean) by Webb (1974a) and as early Rancholabrean by Hulbert (2001). To date, there has been no reliable means of determining a numerical date for that fauna (S.D. Webb, personal communication, June 2002). The two Blancan horn core fragments have been regarded as evidence of an early dispersal of bison across the continent during late Pliocene time (McDonald and Morgan, 1999; Hulbert, 2001). In any event, those specimens must be interpreted within the context of faunal distrubution during the Blancan and have no bearing on the problem of the Irvingtonian/Rancholabrean boundary. At present, the South Carolina specimen (ChM PV6865) from the Ten Mile Hill Beds seems to be the earliest evidence of bison in the Pleistocene of the contiguous United States and thus appears useful in placing the Irvingtonian/Rancholabrean boundary at approximately 0.24 Ma.

* * *

Order PROBOSCIDEA
Family MAMMUTIDAE
Genus *Mammut* Blumenbach, 1799
MAMMUT AMERICANUM (KERR, 1791)
Figures 51, 54a-b

MATERIAL.—ChM PV6077, partial right and left upper third molars, rib fragments, miscellaneous bone fragments.

LOCALITY AND HORIZON.—S.C., Dorchester Co.; bank of Eagle Creek in Tranquil Estates subdivision; Michael Crowell, summer, 1979; Wando Formation, Late Pleistocene.

AGE.— Rancholabrean, Sangamonian.

DISCUSSION.—Remains of the American mastodont are known from many localities in the United States from Blancan through early Holocene times, and during the Rancholabrean this widely distributed proboscidean ranged from Alaska to Florida (Kurtén and Anderson, 1980:344). A number of species and subspecies have been described in the past (see Osborn, 1936), but only one form, *M. americanum*, is recognized in the more recent literature (e.g., Kurtén and Anderson, 1980). Records of mastodonts from South Carolina have been reported by Leidy (1860, as *Mastodon ohioticus*), Hay (1923a, as *Mammut americanum* and as *M. progenium*), Roth and Laerm (1980), and Bentley et al. (1995). The Charleston Museum collection contains 33 teeth of this taxon from South Carolina, most of which were found in the vicinity of Charleston. Though there is no record of the stratigraphic origin of these specimens, the majority of those that are not from Edisto Island probably are from the Wando Formation. However, there is good stratigraphic control for one individual represented in the collection.

During the summer of 1979 a few fragmentary remains of a *M. americanum* skeleton (ChM PV6077) were collected by Michael Crowell from Wando Formation sediments in a spoil pile beside Eagle Creek in the Tranquil Acres subdivision on the

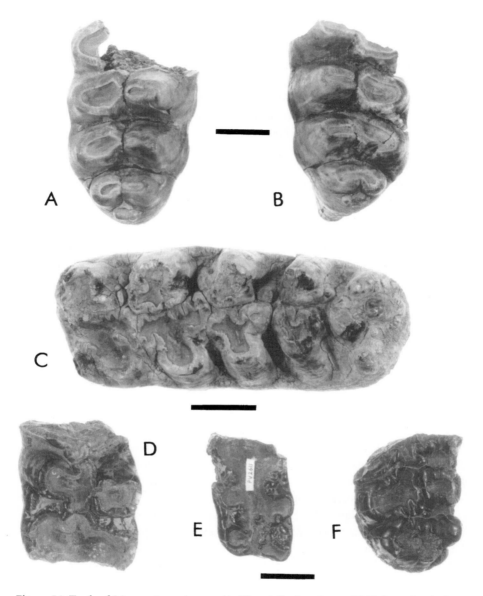

Figure 54. Teeth of *Mammut americanum* (**A, B**) and *Cuvieronius* sp. (**C-F**) from South Carolina. **A,** left M3; **B,** right M3; occlusal views, associated molar teeth (ChM PV6077); Dorchester Co., Wando Formation, (Late Pleistocene); **C,** ChM PV4885, cast of left m3, occlusal view; lingual view; Dorchester Co., Ten Mile Hill Beds (Late Middle Pleistocene). **D,** ChM PV2611, deciduous left m2, occlusal view; **E,** PV2652, partial left M3, occlusal view; **F,** ChM PV5797, right M2, occlusal view; Charleston Co., Wando Fm. (Late Pleistocene). Scale bars = 40 mm.

south side of Ladson Road (County Road 230), approximately 20 miles (32.1 km) north of Charleston. The remains consist only of the left and right M3 (Figure 54a-b), the proximal end of a rib, two limb bone fragments, and miscellaneous small bone fragments. More of the skeleton doubtless was present at one time but was destroyed by a dragline during the chanellization of Eagle Creek. The two molars lack their anterior ends and most of their roots but otherwise are well preserved. Their occlusal surfaces show considerable wear, indicative of an animal well into adulthood.

Family GOMPHOTHERIIDAE
Genus *Cuvieronius* Osborn, 1923
CUVIERONIUS SP.
Figures 51, 54c-f

MATERIAL.—USNM 391982, cast of upper right second molar (cast, ChM PV4894); ChM PV4885, cast of lower left third molar; ChM PV2611, deciduous lower left second molar; ChM PV2652, partial upper left third molar; ChM PV5797, upper right second molar.

LOCALITY AND HORIZON.—USNM 391982, North Carolina, Brunswick Co., pit near Shalotte; Cindy Evans, 1985; upper bed, Waccamaw Formation, Early Pleistocene. ChM PV4885, S.C., Dorchester Co.; drainage ditch behind trailer park circa 0.30 mi. (0.48 km) north of Ladson Road (County Road 230), circa 0..20 mi. (0.32 km) east of Chandler Bridge Creek; Darren Sarine, July 1982; Ten Mile Hill Beds, Late Middle Pleistocene. ChM PV2611, PV2652, S.C., Charleston Co.; Magnolia Phosphate Mine, S.C. Rt. 61 near Runnymede Plantation and Magnolia Gardens; Charles C. Pinckney, Jr., c. 1890; Wando Formation, Late Pleistocene. ChM PV5797, S.C., Charleston Co.; vicinity of Ashley River; Fannie B. Brownfield, donor, 1921; Wando Formation, Late Pleistocene.

AGE.—USNM 391982, Irvingtonian, Pre-Illinoian; ChM PV4885, Rancholabrean, Illinoian; ChM PV2611, PV2652, PV5797, Rancholabrean, Sangamonian.

DISCUSSION.—A member of the *Stegomastodon-Haplomastodon* group of gomphotheres of Pliocene and Pleistocene times in North and South America (Tobien, 1973:243), *Cuvieronius* is known from North American localities primarily in the western United States and in Florida. Remains of this relatively rare proboscidean have been reported from Blancan deposits in Arizona (Kurtén and Anderson, 1980:348) and Florida (Webb and Dudley, 1995), from Irvingtonian beds in Texas (Kurtén and Anderson, 1980:348) and Florida (Webb, 1974a:18, Table 2.1; Webb and Dudley, 1995:648), and from Rancholabrean faunas in Texas (Kurtén and Anderson, 1980:348).

Apparently the first known record of *Cuvieronius* from the early Pleistocene of North Carolina, an upper right second molar of that gomphothere was collected from the Shallotte quarry pit near Shalotte, Brunswick County, North Carolina by Cindy Evans in 1985. The specimen was identified at the National Museum of Natural History but was retained by the collector. However, casts of it are in that institution (USNM 391982) and at The Charleston Museum (ChM PV4894). The tooth evidently came from the Waccamaw Formation, the only Pleistocene unit in the section at the Shallotte pit (personal communication, Robert W. Purdy, April 1997).

Gomphotheres have not been reported previously from South Carolina, the first recognized specimen being a well-preserved lower left third molar (cast, ChM

PV4885, Figure 54b) found in July 1982 by Darren Sarine in the Ten Mile Beds (Late Middle Pleistocene) in the bank of a ditch on the north side of the site of an excavation of the Chandler Bridge Formation (Late Oligocene) conducted by The Charleston Museum from 1970 to 1972 (Sanders, 1980). That site, now covered by a trailer park, is located 0.3 mi (0.5 km) northwest of Ladson Road (County Road 230) and 0.2 mi. (0.3 km) east of the Chandler Bridge canal (32° 57.98' N., 80° 08.96' W., USGS Stallsville 7.5' quadrangle). The specimen was determined as a left m3 of *Cuvieronius* by Clayton E. Ray, and casts were made for the National Museum of Natural History (USNM 336372) and Charleston Museum (ChM PV4885) collections. The collector retained possesion of the specimen.

Three molar teeth (Chm PV2611, PV2652, PV5797) assignable to *Cuvieronius* have been lying unidentified in the Charleston Museum collections for many years but were recently determined by comparison with the aforementioned specimen and from the figures and descriptions in Osborn (1936:568-587). PV2611, a partial juvenile left m2 (Figure 54d), and PV2652, a partial left M3 (Figure 54e), are from the Pinckney collection and thus were apparently found during the Magnolia Phosphate Mine operations. Museum records show that PV5797, a partial right M2 (Figure 54f), was found "in the vicinity of the Ashley River" and was donated by Fannie B. Brownfield in 1921. There is no indication as to when or by whom the latter specimen was collected, but it was probably found during the phosphate mining days, the period in which such specimens were recovered in greater numbers than at any time before 1921. If that were the case, it could have come from any of the several mines situated along the Ashley River. All three specimens are black and display the same degree of wear on the broken surfaces of the root, indicating that they share a similar history of reworking. Thus, they are presumed to have come from the lag deposit at the base of the Wando Formation and therefore to have been reworked from the eroded early Wando sediments of Sangamonian age.

The North Carolina and South Carolina specimens may be referable to *Cuvieronius tropicus*, reported from the Early Pleistocene Leisey Shell Pit of Florida (Webb and Dudley, 1995:648).

SUMMARY AND CONCLUSIONS

The present paper reports 142 specimens representing 37 taxa documenting additional records of *Dasypus bellus, Holmesina septentrionalis, Megalonyx jeffersonii, Eremotherium laurillardi, Canis dirus, Tremarctos floridanus, Arctodus pristinus, Ursus americanus, Smilodon fatalis, Panthera onca augusta, Lynx rufus, Odobenus, Castor canadensis, Neofiber alleni, Neochoerus pinckneyi, Equus* sp., *Tapirus haysii, Tapirus* cf. *T. veroensis, Cervus elaphus, Bison antiquus, Bison* sp., and *Mammut americanum* from South Carolina. Also reported are the first published records of *Pseudorca crassidens, Erethizon dorsatum, Hydrochoerus holmesi, Miracinonyx inexpectatus, Panthera leo atrox, Puma concolor, Erignathus barbatus, Monachus tropicalis, Hemiauchenia* cf. *H. macrocepahala,* and *Cuvieronius* from South Carolina, the first record of *Neofiber* cf. *N. diluvianus* from North Carolina, and the first evidence of *Monachus tropicalis* in Georgia. The holotype of *Alces runnymedensis* Hay, 1923 (ChM PV2551), is redetermined as *Cervalces scotti,* and the stratigraphic origins of the holotype tooth of *Neochoerus pinckneyi* Hay, 1923, and a molar of *Rangifer* cf. *R. tarandus* (ChM PV3411) reported by McDonald et al. (1996) are clarified. All specimens are from strata that are demonstrably of Pleistocene age, except for certain specimens of *Neochoerus pinckneyi* that are of Mid Pliocene age.

Chronostratigraphically, the various taxa are distributed as follows:

Early Pleistocene (Early Irvingtonian)
  Waccamaw Formation, upper bed: *Neofiber* cf. *N. diluvianus, Cuvieronis* sp. [North Carolina], *Miracinonyx inexpectatus, Hydrochoerus holmesi, Neochoerus pinckneyi* (by inference), *Tapirus haysii.*
  Penholoway Formation: *Dasypus bellus, Eremotherium* sp., *Miracinonyx inexpectatus, Neochoerus pinckneyi* (by inference), *Equus* sp., *Cervus elaphus.*
Middle Pleistocene (Late Irvingtonian)
  Ladson Formation: *Megalonyx jeffersonii, Eremotherium laurillardi, Arctodus pristinus, Neochoerus pinckneyi* (by inference), *Equus* sp., *Tapirus* cf. *T. veroensis.*
  Ten Mile Hill Beds: *Dasypus bellus, Holmesina septentrionalis, Odobenus* sp., *Neofiber alleni, Hydrochoerus holmesi, Neochoerus pinckneyi, Equus* sp., *Tapirus haysii, Bison* sp., *Cuvieronius* sp.
Late Pleistocene (Rancholabrean)
  Sangamonian
    Socastee Formation: *Erignathus barbatus, Monachus tropicalis,* (?)*Holmesina septentrionalis,* (?)*Tremarctos floridanus*

Wando Formation: *Megalonyx jeffersonii, Neochoerus pinckneyi, Canis dirus, Arctodus pristinus, Ursus americanus, Odobenus rosmarus, Castor canadensis, Neofiber alleni, Hydrochoerus holmesi, Hemiauchenia* cf. *H. macrocepahala, Rangifer* cf. *R. tarandus, Cervalces scotti, Cuvieronius* sp., *Mammut americanum.*

Wisconsinan

Unnamed unit, Giant Portland Cement Company quarry: *Canis dirus.*

Undetermined offshore unit, Edisto Beach: *Holmesina septentrionalis, Megalonyx jefforsonii, Eremotherium laurillardi, Tremarctos floridanus, Smilodon fatalis, Panthera leo atrox, Panthera onca augusta, Puma concolor, Lynx rufus, Monachus tropicalis, Erithizon dorsatum, Neochoerus pinckneyi, Cervus elaphus, Pseudorca crassidens, Bison antiquus antiquus.*

In summary, 9 taxa are recorded from the Early Pleistocene (Early Irvingtonian) (two from North Carolina and seven from South Carolina), 14 from the Middle Pleistocene (Middle Irvingtonian-Early Rancholabrean), and 30 from the Late Pleistocene (Rancholabrean). The Late Pleistocene records incorporate 17 taxa from the Sangamonian and 16 from the Wisconsinan. One form, *Neochoerus pinckneyi*, previously known only from the Pleistocene, was found to range from mid-Blancan to late Rancholabrean time.

An astragalus (ChM PV6865) found in the Ten Mile Hill beds places bison on the east coast of North America between 240,000 and 200,000 years ago. Presently one of the earliest known records of bison in North America, it provides a suggested boundary of 0.24 Ma for the Irvingtonian/Rancholabrean Land Mammal ages, the beginning of the Rancholabrean having been previously marked by the first appearance of *Bison* in the Pleistocene of North America (Savage, 1951; Lundelius et al., 1987). However, the incomplete record of *Bison* distribution across the continent makes it difficult to gauge the value of any one date as a marker for the earliest appearance of this taxon in the Pleistocene, and other boundaries based on more complete data—perhaps incorporating other taxa as well—will almost certainly be drawn in the future.

SIGNIFICANCE OF THE EDISTO RANCHOLABREAN FAUNA

Fifty-six (39%) of the 142 specimens reported herein are from Edisto Beach, and 15 (40%) of the 37 taxa discussed are represented by specimens from that locality, demonstrating the importance of this locality as a major source of Rancholabrean mammal remains. One of the most productive Pleistocene vertebrate fossil localities on the East Coast of the United States (Hay, 1923a), Edisto Beach has furnished a wide variety of material in the form of disarticulated elements washed ashore by the tide (Hay, 1923a; Roth and Laerm, 1980). Through the years, a sizable assemblage from this locality has accumulated in the collections of The Charleston Museum (Roth and Laerm, 1980). Though far from being a complete representation of the mammals that occurred in the area during the Late Pleistocene, this material is sufficient to demonstrate that the mammalian fauna was composed of forms commonly associated in Rancholabrean local faunas elsewhere, notably in Florida, where Webb (1974a) successfully correlated mammalian faunal chronologies in that state with the Plio-Pleistocene Land Mammal Ages (Blancan, Irvingtonian, and Rancholabrean) defined in faunal sequences in western North America.

Among the Charleston Museum specimens of fossil land mammals reported from Edisto Beach by Ray (1967) and principally by Roth and Laerm (1980) are remains of armadillos (*Glyptotherium*, *Dasypus*, *Holmesina*), ground sloths (*Eremotherium*, *Paramylodon*, *Megalonyx*), lagomorphs (*Sylvilagus*), capybaras (*Neochoerus*), beavers (*Castoroides*, *Castor*), canids (*Canis*, *Urocyon*), bears (*Tremarctos*), raccoons (*Procyon*), jaguars (*Panthera onca augusta*), proboscideans (*Mammuthus*, *Mammut*), perissodactyls (*Equus*, *Tapirus*), camelids (*Palaeolama*), bovids (*Bison*), and cervids (*Odocoileus*). Those taxa, along with sabertooth cats (*Smilodon*), the American lion (*Panthera leo atrox*), pumas (*Puma*), bobcats (*Lynx*), porcupines (*Erethizon*), and elk (*Cervus*) reported in the present paper, adequately demonstrate the Rancholabrean (late Pleistocene) flavor of the Edisto land-mammal assemblage. Specimens representing the Bottlenose Dolphin (Tursiops), the False Killer Whale (*Pseudorca*), the Sperm Whale (*Physeter*), the Gray Seal (Halichoerus), the Monk Seal (*Monachus*,) and the Walrus (*Odobenus*) document a marine-mammal fauna quite separate from the terrestrial fauna and probably of latest Pleistocene or early Holocene age.

The fossil turtle remains from Edisto Beach furnish important clues to the nature of the paleoenvironment. Easily the most abundant fossils found on Edisto Beach, fragments of the shells of freshwater turtles indicate the presence of streams with little or no salinity. The freshwater genera *Pseudemys*, *Trachemys* (sliders), *Kinosternon* (mud turtles), *Chelydra* (snapping turtles), and *Trionyx* (softshell turtles) are well represented in the Charleston Museum material from Edisto Beach. The abundance of their remains suggests that freshwater streams were plentiful, at which time the shoreline would have been a considerable distance seaward of the present strand line at Edisto Beach. Fragments of shells of the large extinct box turtle *Terrapene carolina putnami* Hay are also abundant in the Edisto Beach material, and the presence of this form suggests the nature of environmental conditions in the region prior to post-Wisconsinan submergence. Auffenberg (1958, 1967) and Milstead (1969) have emphasized the apparent value of *T. c. putnami* as an indicator of coastal savanna habitats, to which this subspecies seems to have been restricted. In the higher, forested areas farther inland *T.c. putnami* apparently was replaced by the modern box turtle, *T. c. carolina* (Linnaeus). If that interpretation is correct, the presence of *T. c. putnami* and the absence of *T. c. carolina* in the Edisto material infers a savanna habitat, as might be expected on the flat, gently sloping terrain of the now-submerged margin of the Coastal Plain.

That implication is also consistent with the ecological requirements of the Pleistocene land mammals recorded from Edisto Beach. Two horn cores (ChM PV2393, ChM PV7000) record the presence of *Bison a. antiquus* among the Edisto megafauna. There are other Edisto bison elments in the Charleston Museum collection that represent a form larger than *Bison bison*, the point being that the larger forms (*B. priscus*, *B. latifrons*, *B. antiquus*) are indicative of a paleoenviroment with optimum conditions for feeding. Discussing the marked decline in body size of bison from the giant *B. latifrons* through *B. antiquus* to *B. bison* during late Illinoian-to-late-Wisconsinan time, Gutherie (1984:494) attributed large body size to the presence of a wide variety of plant nutrients available within a long growing season and tied the decline to a reduction of that variety caused by a gradual abbreviation of the annual optimum growing season. The presence of a large bison, *B. antiquus*, among the Edisto fauna thus suggests (1) a sufficient variety and abundance of plants to support populations of larger-sized *Bison* and (2) that those populations, the conditions that they required, and the diversity of other land mammals represented in the Edisto faunal material strongly indicate a corresponding environmental diversity necessary for the survival of all of the coexisting taxa.

Those requirements, coupled with the environmental data furnished by the turtle fauna, suggest a paleoenvironment of low-lying, open savanna grasslands interlaced with freshwater streams and dotted with occasional hammocks of trees and shrubs of sufficient size to support browsing herbivores such as *Eremotherium*, *Mammut*, and *Mammuthus*. In that setting, *Smilodon* and *Panthera leo atrox* were the top predators, probably preying on the smaller ground sloths (*Paramylodon*, *Megalonyx*) and stragglers from the herds of bison, horses, and camelids that roamed the grasslands. *Panthera onca augusta* and *Puma concolor* probably stalked *Neochoerus* and *Tapirus* along the streams and competed with *Canis dirus* for kills of *Odocoileus* in wooded areas.

Certain specimens from Edisto Beach provide evidence of faunal changes associated with glacioeusatic sea level fluctuations. The remains of cetaceans (*Tursiops*, *Physeter* [Roth and Laerm, 1980], and *Pseudorca crassidens*) and pinnipeds (*Halichoerus*, *Monachus*, and *Odobenus*) in the material from Edisto Beach almost certainly date from a time in which the shoreline was not a great distance from its present position. Fossil elements of sea turtles occurring on the beach with the freshwater turtle material furnish additional evidence that the Pleistocene vertebrate remains on Edisto Beach are of mixed stratigraphic origins.

Although there is little direct evidence to indicate the age of the specimens from Edisto Beach, the taxa represented suggest that the Pleistocene land mammals from this locality probably date from Mid-Wisconsinan through Late-Wisconsinan time, since a substantial portion of the early Wisconsinan must be allowed for the gradual development of topsoil to nourish plants that would form suitable habitats on the then-newly-exposed coastal margin as sea level was receding. That concept is supported in part by the aforementioned amino-acid date of 40,000 years obtained from a fragment of shell of a freshwater turtle of the genus *Pseudemys*.

For as long as we contemplate the long-vanished magnificence of a landscape dotted with large spectacular mammals in such abundance as is seen only on the plains of Africa today, the phenomenon of the extinction of the North American Pleistocene megafauna will continue to intrigue us. For some time now, I have wondered about the degree to which the post-Wisconsinan sea level rise may have played a part in the extinction of that fauna. At the Wisconsinan glacial maximum, many thousands of horses, bison, deer, camelids, tapirs, proboscideans, ground sloths, and large carnivores inhabited the broad coastal strip of at least 100 miles (161 km) or more exposed at that time along the Eastern Seaboard. As sea level began to rise with the retreat of the Wisconsin ice sheet, all of those animals had to move gradually inland. While the animals comprising the vanguard of that exodus may have been absorbed into the populations that they encountered further inland, it would seem that, no matter how gradual it may have been, the continuous influx of inland-bound animals would eventually have had some impact upon the populations already occupying the regions along the present-day coastline, perhaps forcing them even further inland and exerting pressure on animals in the Piedmont region. The coastline of South Carolina is approximately 206 miles (331 km) long, and if one allows 100 miles (161 km) for the distance from the present coastline to the edge of the continental shelf (the coastline at the glacial maximum), those figures produce an estimate of 20,600 square miles (53,291 square km) of paleohabitat. The loss of such a vast expanse of feeding grounds along the coast of South Carolina alone, not to mention similar—and at some points even greater—losses along the entire east coast, would seem to have had devastating effects upon some of the megafauna. Grazing animals, such as horses and bison, would have been particularly affected by losses of enormous areas of grasslands. Large herds of those animals moving away from the slowly advancing sea and invading the grazing lands of inland populations might have placed too heavy a burden upon existing food supplies in those areas. Eventually, the Appalachian Mountain chain would have become a formidable barrier to further movement inland except through narrow mountain passes. A continuous influx of indiduals into areas already overgrazed by previous refugees certainly could have caused widespread starvation, and the spread of diseases among greatly weakened animals could have carried off many others.

The foregoing scenario is purely speculative, of course, and certainly has no data to support it. Its purpose is merely to suggest a possible contributing factor to some of the Pleistocene extinctions along the east coast, and to encourage a more rigorous examination of its probablity. The fact that those extinctions occurred at a time when so many thousands of animals were being forced inland by a steadily approaching sea would seem to be more than coincidental.

LITERATURE CITED

Ahearn, M.A. 1981. A Revision of the North American Hydrochoeridae. M.S. Thesis, University of Florida, Gainesville.

Akers, W.H. 1972. Planktonic foraminifera and biostratigraphy of some Neogene formations, northern Florida and Atlantic Coastal Plain. *Tulane Studies in Geology and Paleontology*, 9(1–4):1–139.

Allen, G.M. 1926. Fossil mammals from South Carolina. *Bulletin of the Museum of Comparative Zoology*, 67(14):447–467.

Allen, J.A. 1880. History of North American pinnipeds. Volume 12 in *U.S. Geological and Geographical Survey of theTerritories Miscellaneous Publications*. xvi + 785 pp.

_____. 1887. The West Indian Seal (*Monachus tropicalis*). *Bulletin of the American Museum of Natural History*, 2(1):1–34, pl. 1–4.

Anderson, R.M. 1946. Catalogue of Canadian Recent mammals. *National Museum of Canada Bulletin* No. 102, Biological Series No. 31, v + 238 pp.

Anonymous. 1873. The marl beds and phosphate rocks of South Carolina. Pp. 46–80 *in The Trade and Commerce of the City of Charleston, S.C. from September 1, 1865, to September 1, 1872*. Charleston: Charleston (S.C.) Chamber of Commerce.

_____. 1888. The wonderful phosphate deposits of South Carolina. Industrial Issue, Charleston (S.C.) *News and Courier*.

_____. 1917. Charleston (S.C.) *Evening Post*, 9 March 1917.

Auffenberg, W. 1958. Fossil turtles of the genus *Terrapene* in Florida. *Bulletin of the Florida State Museum*, 3(2):53–92.

_____. 1967. Further notes on fossil box turtles of Florida. Copeia, pp. 319–325.

Bentley, C.C., J.L. Knight, and M.A Knoll. 1995. The mammals of the Ardis Local Fauna (Late Pleistocene), Harleyville, South Carolina. *Brimleyana* No 21:1–35.

Berggren, W.A., F.J. Hilgen, C.G. Langereis, D.V. Kent, J.D. Obradovich, I. Raffi, M.E. Raymo, and N.J. Shackleton. 1995. Late Neogene chronology: New perspectives in high- resolution stratigraphy. Geological Society of America Bulletin, 107(11):1272–1287.

Berta, A. 1985. The status of *Smilodon* in North and South America. *Contributions to Science*, Natural History Museum of Los Angeles County, 379:1–15.

_____. 1995. Fossil carnivores from the Leisey Shell Pits, Hillsborough County, Florida. Pp. 463–499 *in* R.C. Hulbert, G.S. Morgan, and S.D. Webb, (eds.), Paleontology and geology of the Leisey Shell Pits, Early Pleistocene of Florida. *Bulletin of the Florida Museum of Natural History*, 37, Pt II(14).

Blackwelder, B.W. 1981. Late Cenozoic stages and molluscan zones of the U.S. middle Atlantic Coastal Plain. *Journal of Paleontology*, 55(5 supplement):1–34.

Bybell, L.M. 1990. Calcareous nannofossils from Pliocene and Pleistocene deposits in South Carolina. Pp. B1–B9 in *Studies related to the Charleston, South Carolina, earthquake of 1886—Neogene and Quaternary lithostratigraphy and biostratigraphy*. U.S. Geological Survey Professional Paper 1367-A.

Campbell, L., S. Campbell, D. Colquhoun, J. Ernisee, and W. Abbo. 1975. Plio-Pleistocene faunas of the central Carolina Coastal Plain. *Geologic Notes*, 19(3):51–124.

Campbell, M.R., and L.D. Campbell. 1995. Preliminary biostratigraphy and molluscan fauna of the Goose Creek Limestone of eastern South Carolina. *Tulane Studies in Geology and Paleontology*, 27(1–4):53–100.

Cartelle, C., and G. De Iuliis. 1995. *Eremotherium laurillardi*: The Pan-American Late Pleistocene megatheriid sloth. *Journal of Vetebrate Paleontology*, 15:830–841.

Catesby, M. 1731–1743. *The Natural History of Carolina, Florida, and the Bahamas*. London, 2 vols.

Churcher, C.S. 1959. Fossil *Canis* from the tar pits of La Brea, Peru. Science, 130:564–65

_____. 1991. The Status of *Giraffa nebrascensis*, the synonymies of *Cervalces* and *Cervus*, and additional records of *Cervalces scotti*. *Journal of Vertebrate Paleontology*, 11(3):391–397.

_____, P.W. Parmalee, G.L. Bell, and J.P. Lamb. 1989. Caribou from the Late Pleistocene northwestern Alabama. Canadian Journal of Zoology, 67:1210–1216.

Colquhoun, D.J. 1965. *Terrace Sediment Complexes in Central South Carolina*. Columbia: University of South Carolina. 62 pp.

Cooke, C.W. 1936. Geology of the Coastal Plain of South Carolina. *U.S. Geological Survey Bulletin* 867:1–196.

Cope, E.D. 1893. A preliminary report on the vertebrate palentology of the Llano Estacado. Geological Survey of Texas, Annual Report No. 4, 1–136.

_____. 1896. New and little known Mammalia from the Port Kennedy bone deposit. *Proceedings of the Academy of Natural Sciences of Philadelphia*, pt. 2:378–394.

_____. 1899. Vertebrate remains From the Port Kennedy bone deposit. *Journal of the Academy of Natural Sciences of Philadelphia*, 11:193–267.

Cronin, T.M. 1990. Evolution of Neogene and Quaternary marine ostracoda, United States Atlantic Coastal Plain: Evolution and speciation in Ostracoda, IV. Pp. C1-C-43 in *Studies related to the Charleston, South Carolina, earthquake of 1886—Neogene and Quaternary lithostratigraphy and biostratigraphy*. U.S. Geological Survey Professional Paper 1367-C.

_____. L.M. Bybell, R.Z. Poore, B.W. Blackwelder, J.C. Liddicoat, and J.E. Hazel. 1984. Age and correlation of emerged Pliocene and Pleistocene deposits, U.S. Atlantic Coastal Plain. *Palaeogeography, Palaeoclimatology, Palaeoecology*, 47(1–2):21–51.

De Iuliis, G., and C. Cartelle. 1999. A new giant megatheriine ground sloth (Mammalia: Xenarthra: Megatheriidae) from the late Blancan to early Irvingtonian of Florida. Zoological Journal of the Linnean Society 127:495–515.

Downing, K.F., and R.S. White. 1995. The cingulates (Xenarthra) of the Leisey Shell Pit Loca Fauna (Irvingtonian), Hillsborough County, Florida. Pp. 375–396 in R.C. Hulbert, G.S. Morgan, and S.D. Webb, (eds.), Paleontology and Geology of the Leisey Shell Pits, Early Pleistocene of Florida. *Bulletin of the Florida Museum of Natural History*, 37, Pt II(14).

Downs, Theodore, and J.A. White. 1968. A vertebrate faunal succession in superposed sediments from Late Pliocene to Middle Pleistocene in California. *23rd International Geological Congress*, 10:41–47.

Drayton, J. 1802. *A Vew of South Carolina as Regards Her Natural and Civil Concerns.* Charleston. Pp. 252.

Edmund, A. Gordon. 1995. A review of Pleistocene giant armadillos (Mammalia, Xenarthra,Pampatheriidae). Pp. 300–321 in K.M. Stewart and K.L. Seymour, (eds.), *Palaeoecology and Palaeoenvironments of Late Cenozoic Mammals: Tributes to the Career of C.S. (Rufus) Churcher.* Toronto: University of Toronto Press.

Emslie, Steven D. 1995. The fossil record of *Arctodus pristinus* (Ursidae: Tremarctinae) in Florida. Pp. 501–514 in R.C. Hulbert, G.S. Morgan, and S.D. Webb, (eds.), Paleontology and Geology of the Leisey Shell Pits, Early Pleistocene of Florida. *Bulletin of the Florida Museum of Natural History*, 37, Pt II(15).

Evans, P.G.H. 1987. *The Natural History of Whales and Dolphins.* New York; Oxford: Facts on File Press. xvi + 343 pp.

Farlow, J.O., and J. McClain. 1996. A spectacular specimen of the Elk-moose *Cervalces scotti* from Noble County, Indiana, U.S.A. Pp. 322–330 in Stewart, K.M., and K.L. Seymour, (eds.), *Palaeoecology and Palaeoenvironments of Late Cenozoic Mammals: Tributes to the career of C.S. (Rufus) Churcher.* Toronto, Canada: The University of Toronto Press.

Feilden, H.W. 1877. The Post-Tertiary beds of Grinnell Land and North Greenland. *Annals and Magazine of Natural History*, fourth series, 20(120):483–489.

Frazier, M.K. 1977. New records of *Neofiber leonardi* (Rodentia: Cricetidae) and the paleoecology of the genus. *Journal of Mammalogy*, 58(3):368–373.

_____. 1981. A revision of the fossil Erethizontidae of North America. *Bulletin of the Florida State Museum, Biological Sciences*, 27(1):1–76.

Fredén, C. 1975. Subfossil finds of arctic whales and seals in Sweden. *Sveriges Geologiska Undersökning*, Serie C, 710, *Avhandlingar och Uppsatser, Årsbok*, 69(2):1–62.

Gazin, C.L. 1957. Exploration for the remains of giant ground sloths in Panama. *Annual Report of the Board of Regents of the Smithsonian Institution . . . for the year ended June 30 1956*, Publication 4272:341–354.

Gibson, T.G. 1983. Stratigraphy of Miocene through Lower Pleistocene strata of the United States central Atlantic Coastal Plain. Pp. 35–80 *in* C.E. Ray (ed.), Geology and Paleontology of the Lee Creek Mine, North Carolina, I. *Smithsonian Contributions to Paleobiology*, no. 53.

Guilday, J.E., and D.C. Irving. 1967. Extinct Florida Spectacled Bear *Tremarctos floridanus* (Gidley) from Central Tennessee. *Bulletin of the National Speliological Society*, 29:149–162.

Gunter, G. 1947. Sight records of the West Indian Seal, Monachus tropicalis (Gray), from the Texas coast. *Journal of Mammalogy*, 28(3):289–290.

Guthrie, R.D. 1984. Alaskan megabucks, megabulls, and megarams: The issue of Pleistocene gigantism. Pp. 482–510 in Genoways, Hugh H., and Mary R. Dawson, *Contributions in Quaternary Vertebrate Paleontology: A Volume in Memorial to John E. Guilday.* Carnegie Museum of Natural History Special Publication No. 8.

Hall, E.R., and K.R. Kelson. 1959. *The Mammals of North America.* New York: The Ronald Press. 2 volumes. xxx + 1083 + 79 pp.

Harington, C.R. 1977. Marine mammals in the Champlain Sea and the Great Lakes. *Annals of the New York Academy of Science,* 288:508–537.

Harlan, R. 1825. *Fauna Americana: Being a Description of the Mammiferous Animals Inhabiting North America.* Philadelphia: Anthony Finley. x + 318 + 2 unnumbered pp.

Hay, O.P. 1923a. *The Pleistocene of North America and its Vertebrated Animals From the States East of the Misissippi River and From the Canadian Provinces East of Longitude 95°.* Washington, D.C.: Carnegie Institution of Washington. viii + 499 pp.

_____. 1923b. Characteristics of sundry fossil vertebrates. *Pan-American Geologist,* 39:103–104.

_____. 1926. A collection of Pleistocene vertebrates from southwestern Texas. *Proceedings of the U.S. National Museum,* 68, Art. 24:1–18.

Hibbard, C.W. 1955. Notes on the microtine rodents from the Port Kennedy cave deposit. *Proceedings of the Academy of Natural Sciences of Philadelphia,* 107:87–97.

Hibbard, C.W., C.E. Ray, D.E. Savage, D.W. Taylor, and J.E. Guilday. 1965. Quaternary mammals of North America. Pp. 509–525 in H.E. Wright, Jr., and D.G. Frey (eds.), *The Quaternary of the United States.* VII Congress of the International Association for Quaternary Research. Princeton University Press.

Hibbard, C.W., and W.W. Dalquest. 1973. *Proneofiber,* a new genus of vole (Cricetidae: Rodentia) from the Pleistocene Seymour Formation of Texas, and its evolutionary significance. *Journal of Quaternary Research,* 3:269–274.

Holmes, Francis S. 1848. Notes on the Geology of Charleston, South Carolina *Charleston Medical Journal and Review,* 3:655–671.

_____. 1858. *Remains of Domestic Animals Discovered Among Post-Pleiocene Fossils in South Carolina.* James and Williams, Printers, Charleston, South Carolina (private printing). 16 pp.

_____. 1859,1860. *Post-Pleiocene Fossils of South Carolina..* [Pp 1–110, pls. 1–20 (1859); pp. 111–122, pls. 21–28 (1860)]. Charleston, South Carolina: Russell and Jones.

Hulbert, Richard C., Jr. 1995. The giant tapir, *Tapirus haysii,* from Leisey Shell Pit 1A and other Florida Irvingtonian localities. Pp. 515–551 *in* R.C. Hulbert, G.S. Morgan, and S.D. Webb, (eds.), Paleontology and Geology of the Leisey Shell Pits, Early Pleistocene of Florida. *Bulletin of the Florida Museum of Natural History,* 37, Pt II(13).

_____. 1999. Nine million years of *Tapirus* (Mammalia: Perissodactyla) in Florida. *Journal of Vertebrate Paleontology,* 19 (supplement to no. 3, abstracts):53A.

_____. 2001. Mammalia 5: Artiodactyls. Pp. 242–279 *in* R.C. Hulbert (ed.), *The Fossil Vertebrates of Florida.* Gainesville: University Press of Florida.

Hulbert, Richard C., Jr., and Pratt, A.E. 1998. New Pleistocene (Rancholabrean) Vertebrate Faunas From Coastal Georgia. *Journal of Vertebrate Paleontology,* 18(2):412–429.

Jefferson, G. T. 1989. Late Cenozoic tapirs (Mammalia: Perissodactyla) of western North America. Contributions to Science, Natural History Museum of Los Angeles County, 406: 1–21.

Koretsky, I., and A.E. Sanders. In press. Paleontology of the Late Oligocene Ashley and Chandler Bridge Formations of South Carolina. 1. Paleogene Pinniped Bones from the vicinity of Charleston, S.C.: The Oldest known Seal (Carnivora: Phocidae). *Smithsonian Contributions to Paleobiology.*

Krantz, D.E. 1991. A chronology of Pliocene sea level fluctuations, U.S. Atlantic Coastal Plain. *Quaternary Science Reviews*, 10:163–174.

Kurtén, B. 1965. The Pleistocene Felidae of Florida. *Bulletin of the Florida State Museum*, 9(6): 215–273.

_____. 1966. Pleistocene bears of North America. 1. Genus *Tremarctos*, Spectacled Bears. *Acta Zoologica Fennica*, 115:1–120.

_____. 1967. Pleistocene bears of North America. 2. Genus *Arctodus*, Short-faced Bears. *Acta Zoologica Fennica*, 117:1–60.

_____. 1973. Pleistocene jaguars in North America. *Commentationes Biologicae*, 62:1–23.

Kurtén, B., and E. Anderson. 1980. *The Pleistocene Mammals of North America*. New York: Columbia University Press. xvii + 442 pp.

Kurtén, B., and L. Werdelin. 1990. Relationships between North and South American *Smilodon. Journal of Vertebrate Paleontology*, 10(2):158–169.

Leatherwood, S., D.K. Caldwell, and H.E. Winn. 1976. Whales, dolphins and porpoises of the Western North Atlantic. NOAA Technical Report NMFS Circ-396:1–176, 182 figs.

Leidy, J. 1852a. [Reference to a fossil tooth of a tapir]. *Proceedings of the Academy of Natural Sciences of Philadelphia*, 6:106.

_____. 1852b. [Remarks on *Tapirus haysii*]. *Proceedings of the Academy of Natural Sciences of Philadelphia*, 6:148.

_____. 1853. [Notice of] several fossil teeth. *Proceedings of the Academy of Natural Sciences of Philadelphia*, 6:241.

_____. 1854a. Remarks on several fossils indicating new species of extinct Mammalia. *Proceedings of the Academy of Natural Sciences of Philadelphia*, 7:89–90.

_____. 1854b. Notice of some fossil bones discovered by Mr. Francis A. Lincke, in the banks of the Ohio River, Indiana. *Proceedings of the Academy of Natural Sciences of Philadelphia*, 7:199–201.

_____. 1855. A memoir on the extinct sloth tribe of North America. *Smithsonian Contributions to Knowledge*, 7(art. 5):1–68.

_____. 1856. Notice of some remains of extinct vertebrate animals. *Proceedings of the Academy of Natural Sciences of Philadelphia*, 8:163–165.

_____. 1859, 1860. Description of vertebrate fossils. *In* Francis S. Holmes, *Post-Pleiocene fossils of South Carolina*. Pp. 99–110, pls. 15–20 (1859), pp.111–122, pls. 21–28 (1860). Charleston, South Carolina: Russell and Jones.

_____. 1877. Description of vertebrate remains, chiefly from the phosphate beds of South Carolina. *Journal of the Academy of Natural Sciences of Philadelphia*, New Series, 8(3):209–261.

Lundelius, E.L., Jr., T. Downs, E.H. Lindsay, H.A. Semken, R.J. Zakrzewski, C.S. Churcher, C.R. Harington, G.E. Schultz, and S. David Webb. 1987. The North

American Quaternary sequence. Pp. 211–235 in M.O. Woodburne, ed., *Cenozoic Mammals of North America: Geochronology and biostratigraphy.* Berkeley: University of California Press.

McCartan, L., J.P. Owens, B.W. Blackwelder, B.J. Szabo, D.F. Belknap, N. Kriausakul, R.M. Mitterer, and J.F. Wehmiler. 1982. Comparison of amino acid eracemization geochronometry with lithostratigraphy, biostratigraphy, uranium-series coral dating, and magnetostratigraphy in the Atlantic Coastal Plain of the United States. *Quaternary Research,* 18:337–359.

McCartan, L., R.E. Weems, and E.M. Lemon, Jr. 1980. The Wando Formation (Upper Pleistocene) in the Charleston, South Carolina, Area. Pp. A-110-A116 in N.F. Sohl and W.B. Wright (eds.), Changes in Stratigraphic Nomenclature by the U.S. Geological Survey, 1979. *U.S. Geological Survey Bulletin* 1502-A.

McCartan, L., R.E. Weems, and E.M. Lemon, Jr. 1990. Quaternary stratigraphy in the vicinity of Charleston, South Carolina, and its relation to local seismicity and regional tectonism. Pp.. A1–A39 in *Studies related to the Charleston, South Carolina, Earthquake of 1886—Neogene and Quaternary Lithostratigraphy and Biostratigraphy.* U.S. Geological Survey Professional Paper 1367-A.

McDonald, H.G., C.R. Harrington, and G. De Iuliis. 2000. The ground sloth *Megalonyx* from Pleistocene deposits of the Old Crow Basin, Yukon, Canada. *Arctic,* 53(3):213–220.

McDonald, J.N. 1981. *North American Bison: Their Classification and Evolution.* Berkeley, Calif.: University of California Press. 316 pp.

McDonald, J.N., C.E. Ray, and F. Grady. 1996. Pleistocene caribou (*Rangifer tarandus*) in the eastern United States: New records and range extensions. Pp. 407–430 in Stewart, K.M., and K.L. Seymour (eds.), *Palaeoecology and palaeoenvironments of late Cenozoic Mammals: Tributes to the career of C.S. (Rufus) Churcher.* Toronto, Canada: The University of Toronto Press.

McDonald, J.N., and G.S. Morgan. 1999. The appearance of bison in North America. Current *Research in the Pleistocene,* 16:127–129

McDonald, J.N., C.E. Ray, and M.W. Ruddell. 2000. New records and range extensions of *Cervalces, Rangifer,* and *Bootherium* in the Southeastern United States. *Current Research in the Pleistocene,*17:131–133.

McKenna, M.C., and S.K. Bell. 1997. *Classification of Mammals Above the Species Level.* New York: Columbia University Press. 631 pp.

Malde, H.E. 1959. Geology of the Charleston phosphate area, South Carolina. *U.S. Geological Survey Bulletin* 1079:1–105.

Martin, R.A., and S.D. Webb. 1974. Late Pleistocene mammals from the Devil's Den Fauna, Levy County. Pp. 114–145 in S.D. Webb, editor, *Pleistocene Mammals of Florida,* Gainesville, Florida: The University Presses of Florida.

Merriam, J.C., and C. Stock. 1932. The Felidae of Rancho La Brea. *Carnegie Institute of Washington Publication* 422:1–232.

Mills, R. 1825. *Atlas of the State of South Carolina, Made Under the Authority of the Legislature; Prefaced With a Geographical, Statistical, and Historical Map of the State.* Baltimore: F. Lucas, Jr. 56 pp.

Milstead, W.W. 1969. Studies on the evolution of box turtles (Genus Terrapene). *Bulletin of the Florida State Museum*, 14(1):1–113.

Morgan, G.S., and R.C. Hulbert, Jr. 1995. Overview of the geology and vertebrate biochronology of the Leisey Shell Pit Local Fauna, Hillsborough County Florida. Pp. 1–92 in R.C. Hulbert, G.S. Morgan, and S.D. Webb, (eds.), Paleontology and Geology of the Leisey Shell Pits, Early Pleistocene of Florida. *Bulletin of the Florida Museum of Natural History*, 37, Pt I(1).

Morgan, G.S., and J.A. White. 1995. Small mammals (Insectivora, Lagomorpha, and Rodentia) from the Early Pleistocene (Irvingtonian) Leisey Shell Pit Local Fauna, Hillsborough County, Florida. Pp. 397–461 in R.C. Hulbert, G.S. Morgan, and S.D. Webb, (eds.), Paleontology and Geology of the Leisey Shell Pits, Early Pleistocene of Florida. *Bulletin of the Florida Museum of Natural History*, 37, Pt II(13).

Newton, E.T. 1889. Some additions to the vertebrate fauna of the Norfolk «Preglacial Forest Bed» with description of a new species of deer (Cervus rectus). *Geological Magazine*, new series, *Decade* 3,6(4):145–149.

———. 1891. The Vertebrata of the Pliocene Deposits of Britain. *Memoirs of the Geological Survey of the United Kingdom*. xi + 137 pp.

Nowak, R.M. 1979. North American Quaternary *Canis. Monograph of the Museum of Natural History, University of Kansas*, No. 6.

Osborn, H.F. 1923. New subfamily, generic, and specific stages in the evolution of the Proboscidea. *American Museum Novitates*, no. 99:1–4.

———. 1936. *Proboscidea*, vol. 1. xl + 802 pp.

Owens, J.P. 1989. Geologic map of the Cape Fear Region, Florence 1° x 2° Quadrangle and northern half of the Georgetown 1° x 2° Quadrangle, North and South Carolina. United States Geological Survey Map Series, Map 1–1948-A.

Paula Couto, C. de. 1954. Megaterios intertropicals de Pleistoceno. *Anais da Academia Brasileria de Cieñcias*, 26:447–463.

Péwé, T.L., J.A. Westgate, and B.A. Stemper. 1989. Refinement of age interpretation of Quaternary events in Fairbanks area, Alaska. Abstracts, 28 International Geological Congress, 2:602.

Pinsof, J.D. 1991. A cranium of *Bison alaskensis* (Mammalia: Artiodactyla: Bovidae) and Comments on fossil *Bison* diversity in the American Falls area, southeastern Idaho. *Journal of Vertebrate Paleontology*, 11(4):509–514.

Ray, C.E. 1958. Additions to the Pleistocene mammalian fauna from Melbourne, Florida. *Bulletin of the Museum of Comparative Zoology*, 119(7):421–449.

———. 1961. The monk seal in Florida. *Journal of Mammalogy*, 42(1):113.

———. 1965. A glyptodont from South Carolina. *Charleston Museum Leaflet*, no. 27:1–12.

———. 1967. Pleistocene mammals from Ladds, Bartow County, Georgia. *Bulletin of the Georgia Academy of Science*, 25:120–150.

Ray, C.E., F. Reiner, D.E. Sergeant, and C.N. Quesada. 1982. Notes on past and present distribution of the Bearded Seal, Erignathus barbatus, around the North Atlantic Ocean. *Série Zoológica*, 2(23):1–31.

Ray, C.E., and A.E. Sanders. 1984. Pleistocene tapirs in the eastern United States. Pp. 283–315 in H.H. Genoways and M.R. Dawson (eds), *Contributions in Quaternary Vertebrate Paleontology: A Volume in Memorial to John E. Guilday*. Carnegie Museum of Natural History Special Publication No. 8.

Ray, C.E., and A.E. Spiess. 1981. The Bearded Seal, *Erignathus barbatus*, in the Pleistocene of Maine. *Journal of Mammalogy*, 62(2):423–427.

Ray, C.E., A. Wetmore, and D.H. Dunkle. 1968. Fossil vertebrates from the Marine Pleistocene of southeastern Virginia. *Smithsonian Miscellaneous Collections*, 153(3):1–25.

Repenning, C.A. 1987. Biochronology of the microtine rodents of the United States. Pp. 236–268 in M.O. Woodburne, ed., *Cenozoic Mammals of North America: Geochronology and biostratigraphy*. Berkeley, Calif: University of California Press.

_____. 1992. *Allophaiomys* and the age of the Olyor Suite, Krestovka Sections, Yakutia. *U.S. Geological Survey Bulletin*, 2037:1–98.

_____. 1998. North American mammalian dispersal routes: Rapid evolution and dispersal constrain precise biochronology. *Advances in Vertebrate Paleontology and Geochronology*, No. 14:39–78.

Repenning, C.A., and F. Grady. 1988. The microtine rodents of the Cheetah Room fauna, Hamilton Cave, West Virginia, and the spontaneous origin of *Synaptomys*. *U.S. Geological Survey Bulletin* 1853:1–32.

Repenning, C.A., T. R. Weasma, and G.R. Scott. 1995. The Early Pleistocene (Latest Blancan-Earliest Irvingtonian) Froman Ferry fauna and history of the Glenns Ferry Formation, southwestern Idaho. *U.S. Geological Survey Bulletin* 2105:1–86

Richmond, G.M., and D.S. Fullerton. 1986. Introduction to Quaternary glaciations in the United States of America. Pp. 3–10 in Sibrava, V., D.Q. Bowen, and G.M. Richmond (eds.), *Quaternary Glaciations in the Northern Hemisphere,*. Oxford, U.K.: Pergamon Press.

Roth, J.A., and J.A. Laerm. 1980. A Late Pleistocene vertebrate assemblage from Edisto Island, South Carolina. *Brimleyana*, No. 3:1–29.

Sanders, A. E. 1974. A paleontological survey of the Cooper Marl and Santee Limestone near Harleyville, South Carolina (Preliminary Report). *Geologic Notes*, 18(1):4–12.

_____. 1980. Excavation of Oligocene marine fossil beds near Charleston, South Carolina. *National Geographic Society Research Reports*, 12: 601–621.

Savage, D.E. 1951. Late Cenozoic vertebrates of the San Francisco Bay region. *University of California Publications, Bulletin of the Department of Geological Science*, 28(10):215–314.

Savage, J. 1974. The Isthmian link and the evolution of neotropical mammals. Natural History Museum of Los Angeles County *Contributions in Science*, No. 260:1–51.

Seymour, K.L. 1993. Size change in North American Quaternary jaguars. Pp. 343–372 in R.A. Martin and A.D. Barnosky (eds.), *Morphological Change in Quaternary Mammals of North America*. New York: Cambridge University Press.

Simpson, G.G. 1945. Notes on Pleistocene and Recent Tapirs. *Bulletin of the American Museum of Natural History*, 86(2):33–82.

Skinner, M.F., and O.C. Kaisen. 1947. The fossil bison of Alaska and preliminary revision of the genus. *Bulletin of the American Museum of Natural History*, 89:127–256.

Sloan, E. 1908. Catalogue of the mineral localities of South Carolina. South Carolina Geological Survey, Series 4, Bulletin 2. 505 pp.

Spamer, E.E., E. Daeschler, and L.G. Vostreys-Shapiro. 1995. A study of fossil vertebrate types in The Academy of Natural Sciences of Philadelphia. *The Academy of Natural Sciences Special Publication* 16. iv + 434 pp.

Stanley, S.M. 1982. Glacial refrigeration and Neogene regional mass extinction of marine bivalves. Pp. 179–191 in E.M Gallitelli, ed., Paleontology, Essential of Historical Geology, Proceedings of 1st International Meeting, Venice, 1981. Societa Tipografica Editrice Modenese Mucchi, Modena Italy.

_____. 1986. Anatomy of a regional mass extinction: Plio-Pleistocene decimation of the Western Atlantic bivalve fauna. *Palaios*, 1(1):17–36.

Stephenson, L.W. 1912. The Coastal Plain of North Carolina: The Cretaceous, Lafayette, and Quaternary Formations. *Bulletin of the North Carolina Geological Survey*, 3:73–171, 258–290.

Szabo, B.J. 1985. Uranium-series dating of fossil corals from marine sediments of southeastern United States Atlantic Coastal Plain. *Bulletin of the Geological Society of America*, 96(3):398–406.

Tobien, H. 1973. On the evolution of mastodonts (Proboscidea, Mammalia) Part 1: The bunodont trilophodont groups. *Notizblatt des Hessischen Landesamptes für Bodenforschung zu Wiesbaden*, 101:202–276.

True, F.W. 1889. A review of the family Delphinidae. *Bulletin of the U.S. National Museum* 36:1–191.

Tuomey, M. 1848. *Report on the Geology of South Carolina*. Columbia, South Carolina: A.S. Johnson. vi + 293 pp.

Tuomey, M., and F.S. Holmes. 1857. *Pleiocene Fossils of South Carolina*. Russell and Jones, Charleston, South Carolina. xvi + 152 pp.

Van Valkenburgh, B., F. Grady, and B. Kurtén. 1990. The Plio-Pleistocene cheetahlike cat *Miracinonyx inexpectatus* of North America. *Journal of Vertebrate Paleontology*, 10(4):434–454.

Walker, Ernest P. 1975. *Mammals of the World*. Revised 3rd edition by John L. Paradiso. Baltimore and London: The Johns Hopkins Press. 2 vols, xlviii + 1500 pp.

Webb, S.D. 1974a. Chronology of Florida Pleistocene mammals. Pp. 5–31 in S. D. Webb (ed.), *Pleistocene Mammals of Florida*. Gainesville: The University Presses of Florida. x + 270 pp.

_____. 1974b The status of *Smilodon* in the Florida Pleistocene. Pp. 149–153 in S. D. Webb (ed.), *Pleistocene Mammals of Florida*. Gainesville: The University Presses of Florida. x + 270 pp.

_____. 1974c. Pleistocene llamas of Florida, with a brief review of the Lamini. Pp. 170–213 in S. D. Webb (ed.), *Pleistocene Mammals of Florida*. Gainesville: The University Presses of Florida. x + 270 pp.

Webb, S.D., and J.P. Dudley. 1995. Proboscidea from the Leisey Shell Pits, Hillsborough County, Florida. Pp. 645–660 in R.C. Hulbert, G.S. Morgan, and S.D. Webb (eds.), Paleontology and Geology of the Leisey Shell Pits, Early Pleistocene of Florida. *Bulletin of the Florida Museum of Natural History*, 37 Pt II(20).

Weems, R.E., and E.M. Lemon, Jr. 1984a. Geologic map of the Stallsville Quadrangle, Dorchester and Charleston Counties, South Carolina. U.S. Geological Survey Geologic Quadrangle Map GQ-1581.

_____. 1984b. Geologic map of the Mount Holly Quadrangle, Berkeley and Charleston Counties, South Carolina. U.S. Geological Survey Geologic Quadrangle Map GQ-1579.

_____. 1987. Geologic map of the Moncks Corner Quadrangle, Berkeley County, South Carolina. U.S. Geological Survey Geologic Quadrangle Map GQ-1641.

_____. 1988. Geologic map of the Ladson Quadrangle, Berkeley, Charleston, and Dorchester Counties, South Carolina. U.S. Geological Survey Geologic Quadrangle Map GQ-1630.

_____. 1989. Geology of the Bethera, Cordesville, Huger, and Kitteredge Quadrangles, Berkeley County, South Carolina. U.S. Geological Survey Miscellaneous Investigations Series, Map I–1854.

Weems, R.E., E.M. Lemon, Jr., L. McCartan, L.M. Bybell, and A.E. Sanders. 1982. Recognition and formalization of the Pliocene "Goose Creek Phase" in the Charleston, South Carolina, Area. *U.S. Geological Survey Bulletin* 1529-H: H137–H148.

Weems, R.E., E.M. Lemon, Jr., and E.D. Cron. 1985. Detailed sections from auger holes and outcrops in the Bethera, Cordesville, Huger, and Kitteredge Quadrangles, South Carolina. U.S. Geological Survey Open-file Report No. 85–439. viii + 85 pp.

Weems, R.E., E.M. Lemon, Jr., and M. Sandra Nelson. 1997. Geology of the Pringletown, Ridgeville, Summerville, and Summerville Northwest 7.5 minute quadrangles, Berkeley, Charleston, and Dorchester Counties, South Carolina. U.S. Geological Survey Miscellaneous Investigations Series, Map I–2502.

Whitmore, F.C., Jr. 1991. Neogene climatic change and the emergence of the modern whale fauna of the North Atlantic Ocean. Pp. 223–227 in Berta, A., and T.A. Deméré (eds.), Contributions in Marine Mammal Paleontology Honoring Frank C. Whitmore, Jr. *Proceedings of the San Diego Society of Natural History*, 29.

Whitmore, F.C., Jr., and A.E. Sanders. 1977. Review of the Oligocene Cetacea. *Systematic Zoology* (Dec. 1976), 25(4):304–320).

Wilson, M. 1974. The Casper local fauna and its fossil bison; pp. 125–171 in G.C. Frison, The Casper site, a Hell Gap Bison kill on the High Plains. New York: Academic Press.

INDEX

Page numbers with figures are in italics.